Birdsong
for the
Curious
Naturalist

Birdsong *for the* Curious Naturalist

Your Guide to Listening

Accompanied by 734 recordings and over 75 hours of fine listening
on the companion website, BirdsongForTheCurious.com

Donald Kroodsma

HOUGHTON MIFFLIN HARCOURT
BOSTON   NEW YORK   2020

For information about permission to reproduce selections from this book, write to trade.permissions@hmhco.com or to Permissions, Houghton Mifflin Harcourt Publishing Company, 3 Park Avenue, 19th Floor, New York, New York 10016.

hmhbooks.com

Library of Congress Cataloging-in-Publication Data is available.

ISBN 978-1-328-91911-3

Book design by Eugenie S. Delaney

Printed in China

SCP 10 9 8 7 6 5 4 3 2 1

for Janet

The bird did fly
Like the wind in the sky.
The bird did sing
With a special type of ring.

—Jon (third grade)

Thank you for teaching me that if I listen,
there is song everywhere.

—Juliet (kindergarten)

Contents

Acknowledgments   ix

1. Beginnings   1

The magic and mystery of birdsong   1
Where and how to listen   1
Explore on your own   2
Doing science? Yes!   4

2. Birds and Their Sounds   5

Birds!   5
Birds sing and call   6
Birds without a song just call   12
Song or call?   14
Female song and duets   15
Mechanical (nonvocal) sounds   21

3. Why and How Birds Sing   26

Why sing?   26
Courtship songs   30
Singing in the brain   32
Not one but two voice boxes   38
What birds hear   41

4. How a Bird Gets Its Song   44

Inborn songs   44
Improvised songs   47
Learned songs of songbirds, and babbling   49

5. More about Song Learning   56

Big decisions: When, where, and from
   whom to learn   56
Song (and call) matching   58
Song (and call) dialects   63
Mimicry   73

6. Song Learning Often Creates Complex Songs and Large Repertoires 79

Song complexity 79
Small to large repertoires 83
How a repertoire is delivered 94
What? "Songbirds" with no song? 104

7. When to Sing, and How 108

How birds go to roost and awake 108
Energized dawn singing 112
Night singing 123
Singing and calling in flight 130

8. How Songs Change over Space and Time 137

Each individual has its own song 137
Each species has its own song 140
Song changes over evolutionary time,
   from species to species 147

9. Music to Our Ears 153

The music in birdsong 153
More music to our ears 160
Music to my ears—author's choice 164

10. Additional Information 169

EXTRA! EXTRA! WEB BONUS! More birds,
   more sounds! 169
   An invitation to explore, worldwide 170
Who's who? 170
How to hear and see birdsong 178
Recording birdsong 179
Additional resources 179

Notes 180
Photo Credits 183
Index 184

Acknowledgments

I am especially indebted to Olin Sewall Pettingill, Jr., who, during the summer of 1968 at the University of Michigan "Bug Camp" (Pellston, Michigan), asked me to work off a $100 fellowship by recording some birds; that project changed my life. John Wiens, graduate advisor at Oregon State University from 1968 to 1972, gave me free rein to pursue birdsong, as did Peter Marler from 1972 to 1980 at the Rockefeller University Field Research Center.

For this book, a huge thank you to the photographers for sharing their hard-earned images: Wil Hershberger, Laure Wilson Neish, Marie Read, Robert Royse, Brian Small, and John Van de Graaff. And to the generous recordists for sharing their sounds: Mieko Aoki, Greg Budney, Greg Clark, Lang Elliott, Janet Grenzke, Wil Hershberger, Brad Jackson, Richard (Nels) Nelson, and Charlie Walcott.

Thank you to each Birds of North America author, for freely sharing insights about "your" species. My guide to the world of birdsong is written not only with my half century of accumulated knowledge, but also with the wisdom of countless others who share my passions.

Thank you to Eldon Greij, for keen advice on bird photos; to Greg Budney, Karl Fitzke, and Bill McQuay, for your technical skills and contagious enthusiasm for the sounds of our natural world; to Sylvia Halkin and Walter Berry, for your steady support; to Lisa Ekus, for sharing your cardinals; to Mark Robbins, for knowing Kansas bird sounds; to David Stewart, for your inspirational introduction to Australian birds; to graduate students at the University of Massachusetts from 1980 to 2003, for your own inspiring love of birdsong and what it's all about; to Kenda Kroodsma, for your invaluable editorial revisions; and to countless others who have, over the last half century, enhanced my journey with birdsong in so many different ways.

Here's a huge thank you to Bruce Byers, who began as a graduate student in the 1980s, only to become my go-to colleague for any question about birdsong or science in general. Bruce, you're tops—the journey would have been far less fun and insightful without you!

Regina Ryan, you are the best agent ever! Thank you.

Houghton Mifflin Harcourt team: Lisa White, field guide editor par excellence, copyeditor Loma Huh, production editor Beth Burleigh Fuller, designer/compositor Eugenie S. Delaney, and proofreader Fran Pulver. Janet Grenzke studied the entire manuscript, offering insightful suggestions for improvement (she's always right, I've learned). She has been my partner in recording for the past decade, relishing those early mornings, always eager for the next quest. Her fine recordings strengthen the book. Thank you, Janet—for being the queen of logistics, for providing your undying support, for making our book happen.

1. Beginnings

THE MAGIC AND MYSTERY OF BIRDSONG

Birdsong fills our lives with beauty and wonder. The beauty is everywhere and inescapable: a robin caroling after a rain, a thrush fluting in the forest, a wren babbling in the brush, any bird with an unknown name who moves us. Autumn and winter are dreadfully quiet, and we eagerly anticipate the return of spring and song. Singing birds are happy and sing for joy, we want to conclude, but all we can know for sure is our own happiness and joy in hearing them.

And the curious among us wonder, beginning with simple questions that inevitably take us on unanticipated adventures. Just what is a bird? Why does it sing or call? How does it sing, and what does it hear? Where does a bird get its song? Is it inborn or learned? If learned, when and where and from whom does the youngster learn? How do learned songs change from place to place, in the form of dialects? Which birds mimic other species? How many different songs can a bird sing? How do birds go to roost, or awake, or sing so energetically at dawn, or sing through the night? Which birds do what? And in all the listening, what is it about birdsong that we find so beautiful, so musical?

It is the wonder that drives this book, as a quick scan of the Contents will reveal. In the following pages are great joy and happiness, just waiting for you—and adventure, too. Explore the listening projects, and your appreciation and love of birdsong and the natural world will unfurl.

WHERE AND HOW TO LISTEN

Throughout the book, we listen to hundreds of singing and calling birds. You can access the sounds and more information about them in two ways:

1. As you encounter the numbered recordings in the text of this book, go directly to the website BirdsongForTheCurious.com and find the corresponding numbered recording. The text in the book and on the website will guide you in your listening.

2. Alternatively, you can go directly to the recordings on the internet by using the QR ("Quick-Response") codes provided throughout the book. With a QR code reader (free, available for download to smartphones), simply scan the QR code, and you will be whisked to the appropriate page on the website, where you can play the chosen sound and read all about it. Try that for the QR code to the left and you will find yourself at the home page for this book's website.

When you listen, for best effect please use headphones. You minimize extraneous noises this way, and it is then just you and the bird singing directly into your ears. The stereo effect of many of the recordings is also better heard with headphones.

Altogether, more than 75 hours of fine listening and exploring await you. I invite you to linger and listen, to enjoy a sage thrasher singing through the night (p. 127), to marvel at the variety of songs a brown thrasher offers over several hours (p. 92), to appreciate the masterful performance of a western meadowlark (Explore 70). I challenge you to listen to this singing planet as you have never listened before. Don't settle for a few brief sound bites that provide the minimum clues needed to successfully identify a bird to species. No, strive for a deeper understanding of each singing bird, trying to fathom who it is, what's in its head, why in this moment it is singing the way it is. A singing robin is never "just a robin," for example, but an individual expressing his mind, maybe even a "thought." So I pause and I listen, often for hours, and sometimes for days to a single bird, out of admiration, respect, and wonder. The depth and range of recordings in this guide provide you with opportunities that you can find nowhere else, and you can choose to explore as deeply as you wish.

EXPLORE ON YOUR OWN

In 77 places throughout the book, after introducing a topic and providing some examples from common birds throughout North America, I invite you to **Explore** on your own. Although this guide contains a wealth of recorded songs that you can study, there is nothing more satisfying than exploring the natural world firsthand, on your own.

The text for 48 of these **Explore**s is provided in the book, and another 29 are on the website (BirdsongForTheCurious.com).

For some of those explorations you do not need to immediately name the bird you are hearing. It is liberating to be free of naming, and exploring a world without labels can be mind expanding. That idea runs counter, of course, to the primary goal of many birders, which is to pin a label on a bird as quickly as possible and with minimal clues (and then move on). I simply suggest that you first try truly listening, and only second consider the name of what you are listening to. To rephrase one of my favorite book titles, I believe that "Hearing Is Forgetting the Name of the Thing One Hears."

For most invitations to **Explore**, however, finding birds of a particular species is necessary. If you know your birds and where to find them, that's great; perhaps find someone to share your knowledge and fun with. For others, finding a few common birds might be relatively easy, but finding other species may be more challenging. Fortunately, countless bird enthusiasts will be eager to help you. Find a local bird club or Audubon Society, or visit the website of the American Birding Association (http://www.aba.org/). If you are youngish, you are sure to find inspiration and help from the ABA's Young Birders program (http://youngbirders.aba.org/). Ask around and you will discover that a thriving community of bird enthusiasts awaits you.

In all of this listening, I encourage you to engage your eyes as well as your ears, because nothing improves one's hearing more than being able to hear and see a sound simultaneously. "I hear with my eyes," I sometimes say to emphasize the point. Once I *saw* the difference between the songs of an American robin and a scarlet tanager, for example, my ears could never forget. Our eyes are so much better tuned to the world than are our ears, and it is our eyes that can help bring our ears along. A half century ago I began studying sounds with my eyes, but it was a cumbersome, laborious process back then, and it took me about five minutes to print a two-second graph. Now, these graphs (called sonagrams, which I think of as "musical scores for birdsong") scroll by in real time on our computer monitors. I highly recommend that you engage both your ears and your eyes and download Raven Lite, a free program from the Cornell Lab of Ornithology (see "How to Hear and See Birdsong," p. 178).

One final thought about these **Explore** sections: Countless ornithologists have already done extensive research on birdsong, but this book does not explicitly reference all their work. Instead, I refer you to the Birds of North America (BNA) series (https://birdsna.org/), which you can subscribe to for pennies a day or access for free at your local library. There you can find all the latest information on every species mentioned in this book, including references that credit the ornithologists who have done the work. Once you begin exploring a given topic, you will inevitably wonder how your effort aligns with that of others, and the BNA is your entry into this literature. See also Chapter 10, "Additional Resources" (p. 179).

DOING SCIENCE? YES!

Curious naturalists are natural scientists, but then, one might ask, just what is "science"? Here's a pretty standard definition: "Science is the search for truths about the natural world." That's rather abstract, so I'm left wondering how to get started finding those truths. One approach is embodied in this simple statement: "Science is the art of collecting interesting numbers." I like that, because it describes so well the process of exploring (combined with a healthy dose of inspiration and ingenuity, of course!). Scientists collect numbers and count for all kinds of reasons—to understand how often something happens, to help estimate the probability of an event, or simply to describe what we are seeing or hearing.

Pick any species that intrigues you, or any individual that intrigues you, and count something, *anything*. The simplest of questions, such as "How many songs does he sing in a minute?," inevitably leads to more questions, such as "How about the next minute, or another minute at another time of day?," and you will be on your way. How many *dee*s are in the *chick-a-dee-dee-dee* call? How often does that woodpecker or grouse drum, or the snipe winnow? Draw a graph, showing how the event occurs over time, because seeing patterns in the simplest of graphs leads to even more questions.

Question, then describe and explore as you attempt to find an answer; next, refine the question or ask another, based on your best answer. Some of the best science is done in this simple progression, and so much of what you could learn about birdsong in this way would make you the world expert on your chosen species and question, as so little is known about birds. In the **Explore** suggestions, I give you a start, but there are limitless opportunities for you to devise your own counting and listening projects.

2. Birds and Their Sounds

BIRDS!

We know a bird when we see one: It has feathers, flies, and maybe sings. We are not likely to confuse birds with frogs and toads (amphibians), lions and mice and us (mammals), or alligators and lizards and dinosaurs (reptiles). But those reptiles deserve a second thought, because scientists have learned that birds descended from dinosaurs. We now actually think of birds as living dinosaurs with feathers. Technically, in the grand classification of animals, birds are considered to be a special, recently evolved group of reptiles.

The 10,500 species of birds in the world are classified by ornithologists into about 40 major groups, or orders. Some of the common groups in North America include the following, arranged in the most recent "field guide sequence," from those most recently evolved (passerines) to the oldest groups (ducks and geese; read down each column):

passerines	herons, bitterns	cuckoos
("perching birds")	cormorants, anhingas	doves
parrots	shearwaters	grebes
woodpeckers	loons	chickens, grouse
kingfishers	shorebirds, gulls	ducks, geese
owls	rails, cranes	
osprey, hawks	hummingbirds, swifts	
	nightjars	

The passerines, relative newcomers to the flock, are extraordinary. They are only one of many orders yet contain most of the species, about 6,200. Of those, about 4,900 are in a remarkable group called "oscines," or "true songbirds," such as jays and crows, swallows, chickadees, wrens, thrushes, warblers, sparrows, and so many more. These songbirds have brains that guide song learning (see "Singing in the Brain," p. 32) and especially intricate voice boxes that enable complex singing (see "Not One But Two Voice Boxes," p. 38). The other 1,300 passerine species are called "suboscines," or "not true songbirds," which in North America are all flycatchers, such as pewees, kingbirds, and phoebes (in Central and South America are several other suboscine groups: woodcreepers, manakins, cotingas, antbirds, and others). Given their abundance and their abundant singing, the passerines, and especially the songbirds, are the primary focus of this book, but many examples will also come from nonpasserines (see "Who's Who?", p. 170).

Some inevitable confusion about terms is worth addressing head-on here. The thousands of true songbirds (oscines) are classified into a single group *not because they sing* but because of anatomical (e.g., structure of voice box, sperm) and genetic features that they inherited from their common ancestor. These 4,900 (true) songbirds are an evolutionarily cohesive group, and when I use the term "songbird," I refer specifically to this special group. But some of these songbirds do not sing, at least in the usual sense (p. 104), and birds in many other groups certainly sing, such as all of the suboscine flycatchers that we know in North America. So not all songbirds sing, and not all singing birds are songbirds!

With the older groups at the bottom, the most recently evolved groups at the top, the arrangement of 12 representative birds on the opposite page implies an evolutionary tree. Birds near the bottom are from more ancient groups that originated before groups above them.

BIRDS SING AND CALL

After a long winter (the "Big Silence"), our spirits are renewed by the lengthening days and the surge of birdsong in spring. Most often it is the male who rises to the treetops to voice something relatively long and loud and complex, as if he is eager to broadcast his important message far and wide. We hear the beauty and music in his efforts and say he is "singing." (As to why he sings, see "Why Sing?," p. 26.)

We try to distinguish those songs from calls. Calls are typically shorter sounds that are used in specific contexts by both males and females of all ages. A young bird in the nest calls for its parents to feed

Passerines

American robin
(a songbird)

Eastern kingbird
(a suboscine
flycatcher)

Belted
kingfisher

Northern flicker
(woodpecker)

Great
horned owl

Red-tailed
hawk

Ruby-throated
hummingbird

Laughing
gull

Rock pigeon
(common pigeon)

Mexican
whip-poor-will
(nightjar)

Wood duck

Ruffed grouse

it. Always alert, birds call to warn of danger from predators. Foraging birds chatter softly to keep in touch with each other. Songbirds often call high overhead as they migrate during the night. Birds are truly noisy creatures! But a small warning is in order here: ornithologists don't all agree on what constitutes a "song" for many species (see examples in the next section, p. 14).

Here are six songbird examples, with male songs and typical calls.

Black-capped chickadee

SONG: Most black-capped chickadee songs in North America consist of brief, pure whistled notes, *hey-sweetie* (or *fee-bee-ee;* mnemonics in italics), the *sweetie* lower and with a slight waiver midway (♫1). For other versions of the song, see "Song (and Call) Dialects," p. 63.

CALLS: *Chick-a-dee-dee-dee,* the call for which the chickadee is named, is given by both males and females in a variety of circumstances; with predators, the more *dee* notes, the more dangerous the predator (compare ♫2, ♫3, ♫4). Chickadees have a variety of other calls as well, one of which is called "the gargle" (♫5).

For more about the black-capped chickadee, see "Song (and Call) Dialects" (p. 63) and "The Music in Birdsong—Pitch-Shifting" (p. 155).

> ### Explore 1: *Chick-a-dee* calls of North American chickadees.
>
> Chickadees occur in one form or another throughout most of North America (besides the black-capped, the four others most frequently encountered are the Carolina, mountain, chestnut-backed, and boreal chickadees). They are common birds, especially at our winter feeders, and they all *chick-a-dee-dee-dee* in their own style (faster, slower, hoarser, etc.). Get to know chickadees better by counting the number of *dee* notes in their calls under a variety of circumstances. Then maybe try to manipulate the circumstance. Introduce a plastic owl near a feeder, for example, and listen to how the chickadees react. Or introduce a cat into the yard (but, heavens, only briefly!). Use your imagination as to what you think might be of concern to a chickadee, and then

document how they adjust their *chick-a-dee* calls. You can practice your listening and counting with the *chick-a-dee* examples I provide on the website for the book.

White-breasted nuthatch

SONG: The white-breasted nuthatch's song is relatively soft, a rising, somewhat nasal note repeated, *waah-waah-waah-waah-waah,* or faster, *whi-whi-whi-whi-whi-whi,* with each bird apparently singing at two different speeds. In ♪6, hear five slow, then five fast songs. Test your ears to hear how he delivers his slow and his fast songs in these four recordings: ♪7, ♪8, ♪9, ♪10.

CALLS: Amid a chorus of other species, a male and female nuthatch converse intimately in *yank*s and twitters as they forage near one another (♪11).

Sounds of the white-breasted nuthatch vary geographically, a possible clue that this one species could be considered two to four different species. See other examples of this kind of geographic variation within a species in "Each Species Has Its Own Song" (p. 140).

Explore 2: White-breasted nuthatch songs.

A male white-breasted nuthatch is believed to have just two songs, one slow, one fast, and he repeats one song many times before switching to his other (see "How a Repertoire Is Delivered," p. 94). But how many times? Does he favor one song over the other? Just a few minutes of concentrated counting with a nuthatch will take you places where it seems that no one before has explored.

A more advanced project for the curious naturalist might ask if all birds in a population sing at the same slow and fast rates. You will need to record and measure songs, all readily doable if you are interested (see pp. 178–179). I have wanted to do this project for years, as I have suspected that "just two songs per male, one fast and one slow" is oversimplified and might not accurately reflect what nuthatches actually do (see also p. 169, on the red-breasted nuthatch). This listening project is good for late winter or early spring, as nuthatches have an especially early singing season.

Veery

SONG: The male veery sings in a loud, breezy, downward spiral, *da-veeyur, veeyur, veeer, veer,* echoing as if he sings into a metal pipe (♩12).

CALL: Simple *veeeer* notes are slurred downward, given here by a small flock of migrants descending from a night of flying during early September (♩13). About an hour before sunrise, veeries awake on the nesting grounds and immediately engage in a calling chorus, before they begin singing (see "How Birds Go to Roost and Awake," p. 108).

American robin

SONG: It's "the all-American Robin," singing several low caroled phrases, *cherrily, cheer-up, cheerio,* followed when excited by a high, screeching *hisselly* (or *eek!*), a brief pause, then continuing. How fascinating to listen how he chooses to combine the two phrase types, with no two sessions ever alike—I never tire of listening. Try these three examples: ♩14, ♩15, ♩16.

SONG WITHOUT THE CAROLS, ONLY *HISSELLY*S: (♩17)What? For what purpose? Why leave out the carols? This robin sang quietly in the top of a tree, with a second robin nearby (the second bird can be heard calling at 0:11, 0:14, and more). What was the mood of this *hisselly* bird? He seemed agitated, highly conflicted with that second robin so near.

CALLS: Robins are so expressive!

A host of robins settling into a winter roost, with howling wind and hooting great horned owl (♩18).

Agitated calling when a nest is threatened by a human (♩19).

A robin sounding not at all pleased as an adult, predatory northern hawk owl perches above (♩20).

A "hawk alarm," of the kind given by many songbirds when a hawk is sighted, so high and thin that the hawk has trouble hearing and locating the calling bird (♩21).

More about robins: "How a Repertoire Is Delivered" (p. 94).

Explore 3: American robin songs.

Discover something about robins that no one else seems to have explored. A robin sings "several" low, caroled phrases in a series before pausing or offering a high, screechy note (a *hisselly,* though some hear it as *eek*), but exactly how many carols does he sing? Or how many *eek*s? Do the numbers change through a singing session, or from early morning to midmorning to midday to afternoon to evening? Or from one week or month to the next? And how about the ratio of carols to *eek*s? What might you learn about the mind of a robin by simply counting his songs like this? No one yet knows.

You can listen to patterns in robin singing anywhere, thereby documenting a general pattern in robin singing for whatever time scale you choose. Even more satisfying, however, might be to learn how an individual robin sings over time. Many robins can be recognized as individuals by a unique carol or two (see "How a Repertoire Is Delivered," p. 94), and you can use those unique carols to ensure that you are listening to the same robin from hour to hour or day to day.

Gray catbird

SONG: The male gray catbird offers a seemingly endless variety of squeaks, whines, rattles, whistles, and gurgles, about 90 per minute (♫22).

CALL: A simple cat-like *meow,* the call that gave the bird its name (♫23).
More about catbirds: "Improvised Songs" (p. 47).

Common yellowthroat

SONG: The common yellowthroat's song is a bright, rolling, well-enunciated *wichity-wichity-wichity,* typically a three-syllable phrase repeated several times (♫24).

CALL: A split-second, single note, low and husky, *tchep,* gives away the yellowthroat in the tangles (♫25); less often one hears a chatter, suggesting aggressive intentions (♫26).

More about yellowthroats: "Big Decisions: When, Where, and from Whom to Learn" (p. 56) and "Song (and Call) Dialects" (p. 63).

BIRDS WITHOUT A SONG JUST CALL

Some birds have nothing that resembles what we would consider a song, and instead use only calls. That said, an important reminder is necessary here: The categories of "song" and "call" are only *our best human attempt* to place bird sounds into two bins that make sense to us (see also "What? 'Songbirds' with No Song?," p. 104). For some species, the distinction between songs and calls is blurred—a few examples are provided on pages 14–15.

Canada goose

HONKING. The call of a male Canada goose is a low *honk,* that of the female a higher *hrink,* with the highly coordinated honking of a pair often sounding like one bird.

Their two voices are most easily recognized when the birds are well separated from each other, as in this pair of geese that I surprised on a lake in northern Michigan (♩27).

Try to identify his and her contributions in a pair taking flight from a sandy beach (♩28) or, for more of a challenge, in a prebreeding flock on a pond in spring (♩29).

Mallard

The quintessential *QUACK!* of the female mallard is sometimes given singly, sometimes continuously for minutes on end (♩30), and sometimes as a decrescendo (a series that grows softer in volume). The male does not quack. His call is far quieter, a reedy, rasping *rab* or double *rabrab,* heard here from a small flock on a winter pond (♩31).

> **Explore 4: Mallard *quacks* and *rabrab*s.**
>
> (See website for details.)

Chimney swift

Swifts twitter and chitter on the wing as they forage for insects overhead. They seem especially active in the hour before going to roost—in the chimney, of course (♫32).

> ### Explore 5: When and why do chimney swifts twitter?
>
> Just when do chimney swifts twitter? When they are flying all alone, or flying with others, or when they meet other swifts on a flyby? Learning *when* they call will provide hints as to *why* they call. As with so many explorations I suggest in this book, you are on your own, as I can find no evidence that anyone has been curious enough to ask, much less answer, these simple questions of chimney swifts.

Killdeer

Killdeer are most often heard as *kill-dee,* or *dee-ee,* sounding upset at our presence (♫33). In courtship, the male flies about his territory, announcing his name, *killdeer, killdeer, killdeer, killdeer* (♫34). He seems to be proclaiming his ownership of a territory and attempting to impress a female, two functions that we readily attribute to "song" in other groups of birds, yet we don't normally think of killdeer as singing. Perhaps it is the simplicity of the *killdeer* sound that biases us to label it a call; in contrast, many shorebirds in the Arctic have far more complex flight vocalizations that we comfortably label "songs." As for me, I hear a male killdeer *singing* in flight!

Double-crested cormorant

At dawn from their overnight roost, these cormorants seem to cough and sneeze, snarl and moan, often grunting like pigs (♫35). Never does one hear anything that we would consider a "song."

Red-shouldered hawk

Just out of the nest, a young red-shouldered hawk tests its wings, calling continuously as it soars above the treetops, *kee-aah, kee-aah, kee-aah, kee-aah, kee-aah, kee-aah* (♪36). That youngster already calls much like an adult, as heard in ♪37. In fact, stake out a nest and follow the development of the nestlings and you will hear that they increasingly sound like adults as they mature; in ♪38, the younger birds still in the nest barely manage a recognizable *kee-aah,* while an older sibling perched nearby already sounds very adultlike.

Explore 6: Red-shouldered hawk counting games.

Enjoy counting? If you live in red-shouldered hawk country, count the number of *kee-aah* phrases in a single burst from one of these hawks. Just one tally can lead to questions. How many phrases are in the next series? And the next? How might these *kee-aah* outbursts vary through the day, or with the season, or when two hawks seem to answer each other? Given your close attention, you may discover other ways in which the birds vary their calling. Such information would be indispensable when trying to understand what the hawks are saying, and under what conditions, yet no one, not even professional ornithologists, has addressed these questions.

SONG OR CALL?

For most species, especially the flycatchers and the songbirds (i.e., the passerines), we can feel reasonably comfortable identifying "songs" and "calls." But in other groups, these two categories of vocalizations are less clear, largely because we humans have imposed on the birds a simplistic, two-category classification based on an incomplete understanding of the function of the sounds. It is best to admit all that up front, and be reminded of it repeatedly.

Here are just three examples in which the definitions of song and call become blurred.

Red-bellied woodpecker

Song or call (♪39)? With the female visiting twice in this sequence, a male red-bellied woodpecker perches beside the nest cavity, vocalizing repeatedly in much the way a male songbird sings. These nearly 21 minutes are part of a much longer series from this male.

American wigeon

Song or call? A male wigeon swims about as if patrolling a territory, repeatedly uttering a three-parted, airy whistle, *wee WHEE whew,* the first *wee* note barely audible over the river noise (♪40). I don't usually think of "singing ducks," but this vocalization seems to have all the characteristics of a "song."

Pied-billed grebe

Song or call? Both the male and female pied-billed grebe give an extended series of *kuk kuk kuk* notes that transition into a series of *kaow kaow kaow* notes (♪41), the male often extending this sound with donkeylike braying. An unpaired male in spring sounds off repeatedly, as if seeking a mate, in a

behavior similar to that of an unpaired male songbird singing to attract a mate. It would therefore follow that these grebes "sing," at least by the usual definition applied to songbirds.

FEMALE SONG AND DUETS

In all species, parents must communicate with each other, especially when raising a family. Among birds, the interactions can take a number of forms. Quiet, simple calls may suffice, but often the exchanges are more conspicuous. In some tropical species, especially wrens, the coordinated singing duet between male and female is so precise that you cannot know they are two birds unless you stand between them. In other species the duet is more loosely coordinated as male and female respond to each other with various songs, calls, or drums. Complex greeting or courting ceremonies occur in many species as well. Here are just a few of my favorite examples.

Sandhill crane

Sandhill cranes trumpet, bugle, rattle, or croak (take your pick), and *loudly*, with authority. The male's call is similar to but noticeably lower than his mate's, and she often follows him so quickly that the two sound like one bird.

♪42: During early spring, a migrant pair vociferously establishes its territory. Seventy yards away, across a still-frozen lake, two cranes declare that they are paired and that they own this real estate. In the first segment (0:00–1:33), hear how her first three calls are similar to his but on a noticeably higher frequency; then, at 0:08, she begins more animated calling, with two to three outbursts for each one of her mate's. In the second segment (1:33–3:43), the cranes take flight, circling about the lake. With your eyes closed, stereo headphones on, hear how the male flies off to the right, followed by the female; she returns first, followed by him, and they both fly far off to the left, still above the small lake. They soon return, crossing in front of us, and fade into the distance to the right.

♪43: A lone female intruder elicits a strong response from the resident pair! She arrives from the left (south), calling faintly in the distance at first, but when the resident pair hear her, they begin to respond (at 0:29) from the adjacent woods where they have been foraging. At 1:27, the intruder calls intensely. By 3:17, the resident pair is lakeside and much louder, with the intruder again calling intensely at 3:21. Five seconds later (3:26), the resident female takes flight, silently chasing the intruding female off the lake to the right. The male now calls by himself (3:37 to 3:58), then flies to meet her upon her return flight, their duets resuming in flight at 3:58, and continuing when they land together at about 4:35. Vigorous duetting on the lake shore continues until 22:00, when both cranes take flight, continuing their duets as they fade into the distance to the right.

♪44: Immediately after mating, when the male flaps down off her back to the water (heard at 0:2.5), she calls loudly (0:5.7), he following within a third of a second and overlapping her. Note that this is the same intense female call that the intruding female used in ♫43 (e.g., at 1:27).

♫45: Two California cranes foraging in a marsh. They call with less intensity than did the spring pair establishing its territory in Michigan.

♫46: Listen to the din of this migrating flock on the North Platte River in Nebraska during March and identify the duets between mates (four noticeable duets in the first ten seconds).

Sooty shearwater

These shearwaters are largely silent during the non-breeding season when we see them soaring over the North Atlantic or Pacific Oceans. But visit them where they breed, such as in the Falkland Islands, and listen to the seductive *oooooooooh*ing and *aaaaaaaaaah*ing of male and female as they greet each other deep in their nesting burrow (♫47).

Anhinga

In a greeting ceremony at the nest, both male and female anhinga arch and sway their necks like two dancing snakes. With bills open, their undulating clicking and chattering *chitter-chitter-chitter-chee-cheer-chitter-chitter-chitter* accompany the dance (♫48).

Carolina wren

During heated exchanges with other territorial pairs, the female Carolina wren often chatters loudly when her mate sings (♫49), as if together they coordinate defense of their territory. She

also chatters during more intimate exchanges with her mate, such as when they leave the roost in the morning (♫50).

More about the Carolina wren: "Singing in the Brain" (p. 32) and "Learned Songs of Songbirds, and Babbling" (p. 49).

Wrentit

In the Pacific coastal scrub, the male wrentit's bouncing-ball song is unmistakable: *pit-pit-pit-pit tr-r-r-r-r-r-r-r-r-r-r-r-r* (♫51 and ♫52, from two different males; note their distinctive voices; see also "Each Individual Has Its Own Song," p. 137). The male's song is often all that you hear, but keep your ears peeled, and on occasion you will hear the female respond with her simpler but prolonged *pit-pit-pit-pit-pit-pit-pit,* without the trilled ending (♫53).

> **Explore 7: Duets by Carolina wrens and wrentits.**
>
> (See website for details.)

Baltimore oriole

The female Baltimore oriole sings from the nest while incubating, typically in response to her mate's songs nearby (♫54). In this example, excerpted from several hours one May morning, the male's complete song is relatively simple and consists of four notes, *tew tew twe twe,* two low notes followed by two higher notes, but he often offers only the first one or two notes. She responds, sometimes with a single note, but often with an entire song, a wonderfully rhythmic *we-chew-dle-wee-dle-wee-dle-weet.*

Red-winged blackbird

From any elevated perch in his marsh territory, the male red-winged blackbird spreads his tail and shows off his brilliant red epaulets, delivering a loud, gurgling *konk-la-reeeeeeee* (♫55). But he is not alone, as one or more of the females nesting in his territory may respond to him immediately with a harsh chatter, *chit-chit-chit-chit-chit* (a female responds five times in the first 45 seconds of ♪56). Females also have another commonly used call (song?) as well, though it typically is not used in response to the male's song. It is a grating, descending *teer-teer-teer-teer,* believed to be an aggressive signal toward other females (heard at 0:51 in ♪56), and is sometimes given in combination with the *chit* call (e.g., at 5:51 and 6:44).

More about red-winged blackbirds: "Song (and Call) Dialects" (p. 63) and "Energized Dawn Singing" (p. 112).

Explore 8: Listening to female red-winged blackbirds.

Red-winged blackbirds offer excellent listening opportunities. Get comfortable beside a local marsh where red-wings nest. Listen to the male sing, but then listen for the response from nearby female. Locate one female, perhaps foraging at water's edge, or maybe where she is building a nest or even incubating her eggs. Then listen intently for exactly when she responds to his singing. Is she more responsive after he has been quiet for a while? Is she most responsive during his first few songs, but then not so much after that, or when he changes the song he has been singing? (See "How a Repertoire Is Delivered," p. 94.) Does the probability of a response vary with time of day? What numbers could you collect to better understand the female's view of her world? Again, you are entering largely uncharted territory. Enjoy the exploring!

Northern cardinal

During early spring, before nesting, the female cardinal often sings from the treetops, as does the male; one must look to know who is singing. She also leaves the night roost with sharp *chip* notes, then sings, just like the male (♪57); in this example, she chooses a song that is indistinguishable from the song of her mate, who left his roost earlier and now sings some distance away (compare her songs before 1:33 with his after 1:33 in this recording). In

a second example (♫58), the female again calls and sings, this time also matching the song of her mate (female songs are before 1:23, male songs after).

Once the nest is built and the eggs are laid, she often sings from the nest as she incubates the eggs, usually in response to her mate singing nearby. In ♫59 and ♫60 listen to 17 such examples; in 8 of those examples, the female responds to the male with the same song that he is singing. During another 6 examples she responds with a nonmatching song; in the remaining 3 examples, the male could not be heard. There is nothing second-rate about her songs. They are every bit as fine as the male's.

More about cardinals: "Not One But Two Voice Boxes" (p. 38), "Song (and Call) Matching" (p. 58), "Small to Large Repertoires" (p. 83), and "How Birds Go to Roost and Awake" (p.108).

Rose-breasted grosbeak

The male rose-breasted grosbeak is an *operatic* robin, it is said, a robin that's had voice lessons, delivering with seemingly great enthusiasm a spirited ramble of rich, melodious, slurred whistles (♫61; for comparison, an American robin can be heard in the background).

But the female sings, too, heard here as she incubates the eggs in the nest (♫62). Hear her relatively weak song at 0:03, 0:16, and again at 0:56, accompanied by five *chip* calls from 0:29 to 0:56 (listen with stereo headphones). While she is calling and singing, the male is nearby, as he is returning to take his turn at incubation. He is heard softly in the distance at first, then loudly off to the left at 1:17, then settled on the nest by 1:30. He now sings lustily as he incubates, departing a little after 9:00 when the female returns. At 9:11 he sings just off to the right of the nest.

More about rose-breasted grosbeaks: "Courtship Songs" (p. 30).

MECHANICAL (NONVOCAL) SOUNDS

Songs and calls are vocal, produced by a bird's voice boxes (syringes—singular, syrinx; see p. 38), but birds also communicate with mechanical sounds. Many species use their wings. Some stomp their feet. Woodpeckers are master percussionists, drumming by slamming their bills into tree trunks or other substrates.

Mourning dove

Listen as a mourning dove arrives on whistling wings, perches briefly, then takes flight again, each flap of its wings registering a brief whistled sound up around 2000 Hz, about two octaves above middle C on a piano (♩63). The pulsed whistles tell us that wingbeats are 6 to 8 per second when landing, 11 when accelerating on takeoff, then slowing to 6 for normal flight. The function of these whistling wings, made by both males and females, is unknown.

More about mourning doves: "Inborn Songs" (p. 44).

> **Explore 10: Whistling wings of mourning doves.**
>
> Mourning doves offer listening opportunities everywhere across North America. Do they always whistle with their wings when taking off and landing? How about when they approach or leave their nest, when you would expect them to be silent? Or maybe the wings whistle only when we spook them, or when they are spooked by something else? Are all takeoffs identical, with the same duration of whistling? The official Birds of North America account declares "Function [of wing whistling] unknown but may have some alarm-sounding value at takeoff." But, I then ask, what about the function at landing?

Ruddy duck

During what ornithologists call a "bubbling display," the male ruddy duck first erects his head, tail, and crown feathers while inflating the neck feathers with air. With his bill he then beats his neck, bubbling air into the water. The beating intensifies to the finale, as the tail arches over the back and the head is extended, all of this ending with a low, belching, vocal *waaaaaaa* (♩64).

Common nighthawk

Erratically the nighthawk flies about overhead, seemingly bouncing on air, calling, *peent . . . peent . . . peent*. And then he dives, extending his wings down and forward, the air briefly rushing through the extended feathers sounding like a miniature freight train, *VROOOM* (♩65).

Explore 11: *Peent*ing and *VROOM*ing common nighthawks.

I wonder, how many times does he *peent* before each display dive? How might that pattern change over time? The low-frequency *VROOOM* carries great distances from neighboring territories, so I could also hear if a male chooses to *VROOOM* just after one of his neighbors, or if they coordinate or adjust the timing of their booming display dives in any way.

Broad-tailed hummingbird

The broad-tailed hummingbird trills with his wings. With each wing stroke, one of his tapered, outer wing feathers vibrates, producing a pulse of sound about 40 times each second during normal flight, and up to 52 times when hovering (two examples: ♩66, ♩67).

Wilson's snipe

Circling high above the ground territory, the male (and sometimes the female) snipe power-dives, directing rushing air from each wing stroke through special extended outer tail feathers. Each wingbeat generates a brief *wu*, so that the entire five-second display dive produces an extended, winnowing

wuwuwuwuwuwuwuwuwuwuwuwuwuwuwuwuwuwu,
with 35 to 40 *wu*s revealing as many wingbeats during the dive (♩68).

Explore 12: Winnows of Wilson's snipe.

It is great fun to follow a snipe high overhead as he circles about, from one swooping dive to the next, seemingly intent on stitching heaven to earth. Once airborne, how long does he remain aloft, and how many display dives does he complete? Does he have favorite places to dive? Is there any effort to coordinate his display dive with those of a neighbor? So many questions, but as yet no answers.

Northern flicker

Woodpeckers drum, slamming their heads (by way of the bill, of course) into substrates in a way that would scramble our human brains. Woodpecker brains, however, are spared by several shock-absorbing mechanisms, such as strong neck muscles, spongy helmetlike layers to their skull, and "tongues" that wrap around and cushion the skull.

And why don't their eyes pop out of their sockets with each whack of a tree? Like us, woodpeckers have an upper and lower eyelid, but they also possess a third horizontal eyelid that closes just before impact, securing and protecting the eye. In this selection (♩69), the flicker drums five times at the rate of about 22 beats per second, and also offers his loud *wick-wick-wick* song (at 0:36).

Pileated woodpecker

The pileated's deep, resonant drumroll is distinctive, peaking midway at about 16 beats per second (♫70). Typically from high on the bole of a dead tree, the drumming reverberates far and wide throughout the forest. Females also drum, though less frequently than do males. In the background of this second recording (♫71), hear a nearby ruffed grouse drumming with his wings (first heard at 4:22, then about once each minute thereafter; more about the drumming grouse in "Each Species Has Its Own Song," p. 140).

More about the pileated woodpecker: "How Birds Go to Roost and Awake" (p. 108).

The Sapsuckers

Yellow-bellied sapsucker
Red-naped sapsucker
Red-breasted sapsucker

Slow and irregular, sapsucker drums are unlike those of any other woodpeckers. They begin with an introductory roll of several rapid strikes over about a quarter second, then slow to an uneven cadence of curious double strikes, about four to six per second, depending no doubt on the species, the individual, and his mood.

Listen first to a yellow-bellied sapsucker drumming for several minutes during early spring (♫72). Females also drum, but not so often, and when they do, the drum is softer and briefer (some of the weaker drums in ♫73 are from a female, e.g., at 0:02). In this chorus of dawn drummers (♫74), beginning 15 minutes before sunrise, listen to how these sapsuckers respond to each other and to the variety of substrates they use, especially beginning at 20 minutes into this performance.

Of all woodpeckers, sapsuckers seem most likely to find some man-made resonant structure to use instead of the standard tree trunk. Here are two tool-using sapsuckers, one drumming on a metal fire tower (♫75), another on a large metal bell (♫76). House gutters are also favorites.

The drummings of the red-naped (♫77) and red-breasted (♫78) sapsuckers have the same general pattern, an introductory roll followed by the irregular delivery of double strikes.

| **Explore 13: Seeing and hearing the "double strike" of sapsuckers.**

(See website for details.)

3. Why and How Birds Sing

WHY SING?

The consensus is that birds sing to defend a territory and attract a mate. Let's think about that. Outside of the nesting season, perhaps after migrating to Central or South America, both males and females of many species defend territories with simple calls, not with songs. Songs don't seem to be all that crucial for defending a territory. When those migrants return to breed in North America, the males sing, suggesting that their songs are used primarily to attract or impress females. Songs might also be used in contests with other males so that females can in some way judge the quality of a male singer as a breeding partner. "The song is for her" is the best educated guess as to why male birds sing so much.

A territorial bachelor seeking a mate will sing all day long. Once paired, he sings far less, often then restricting his singing to an intense effort during the dawn chorus (p. 112). Why does he continue to sing then if he already has his mate? Probably because he is attempting to impress not only his female but also those on neighboring territories, as DNA fingerprinting has revealed that females may choose to mate not only with their own partner but with other males on nearby territories. We thus speak of a male indigo bunting, for example, as being "socially monogamous": He is paired with one female, and together they tend their offspring, but a male who is especially attractive to females in a neighborhood might impregnate several of them as well. He apparently sings for those females, too.

Here are six examples of bachelors singing throughout the day.

Eastern whip-poor-will, Mexican whip-poor-will

The eastern whip-poor-will sings his name, *whip-poor-will, whip-poor-will, whip-poor-will,* about once each second, relentlessly "jarring the night" (hence the name of the subfamily they belong to, "nightjar"). From an unpaired male seeking a mate under a full moon, I once counted nearly 21,000 *whip-poor-will* songs in one night. Here are two examples, both from Kentucky: ♪79, ♪80.

In southern Arizona and New Mexico, the closely related Mexican whip-poor-will sings in much the same way, with a burry twist to the song (♪81). The differences in the presumably innate songs of the eastern whip-poor-will and this whip-poor-will of the Southwest was, in fact, one of the primary arguments for elevating the southwestern birds to full species status in 2010. (See also "Song Changes over Evolutionary Time, from Species to Species," p. 147.)

Virginia rail, sora

Kidick . . . Kidick . . . Kidick, or heard as *Tick-it . . . Tick-it . . . Tick-it,* the male Virginia rail sings from the marsh all day long in early spring, and will continue to do so until a female pairs with him (♪82). If he is unsuccessful, he gradually gives up on attracting a female sometime during the nesting season, probably when he innately calculates that his prospects are approaching zero. Another male rail in the marsh, the sora, repeatedly calls its name "*so-RAH*" until it pairs (♪83).

Brown thrasher

Unpaired during early spring, the male brown thrasher sings nonstop all day long from the tree-tops, a bewildering variety of paired doublets, his message clearly *I'm here I'm here, lookit me lookit me, settle here settle here, good place good place* . . . The moment a female joins him on his territory, the extended singing sessions abruptly end, again suggesting that the singing effort is primarily for her, to attract her to his territory (♪84).

More about the brown thrasher: "Song (and Call) Matching" (p. 58), "Small to Large Repertoires" (p. 83), and "The Music in Birdsong—Theme and Variations" (p. 158).

Black-throated blue warbler

The male black-throated blue warbler sings a lazy, rising *zur zur z-z-zreeee,* or *I'm so la-la-zeeeeeee* (♪85). A bachelor male who has just arrived in spring can sing thousands of these songs from before sunrise until late afternoon, when it seems that he chooses to give up for the day and hopes for better success the next. This extraordinary effort ends only when *she* arrives (and about then a whole new pattern of singing begins at dawn, with sporadic singing through the day; see "Energized Dawn Singing," of yellow warblers, for example, p. 115). Should she happen to disappear, he resumes his day-long singing pattern.

Explore 14: Indefatigable, bachelor males advertising for a mate.

For a week or two during early spring, get to know a singing male by documenting his singing effort to attract a mate to his territory. Males typically return from migration before females; each male establishes his territory and then, until a female joins him, he is a bachelor and typically sings all day long. Enjoy all-nighters? Find a local nightjar. Don't limit yourself to whip-poor-wills, but consider their relatives, such as a chuck-will's-widow or a common poorwill. Or choose a songbird just returning from migration; warblers are fascinating, because many species use different songs in advertising for females and in fighting with males (e.g., p. 115). If you don't want to choose a particular species, you could document the decline in singing for all species in an area by simply counting the number of songs or singing birds that can be heard during a selected time period each morning (e.g., a 15-minute block of time each day beginning at 9 A.M.).

Among some species, each male attempts to attract several mates (a mating system known as "polygyny"), and as a result his singing (or "displaying") can continue unabated throughout the nesting season. He is, in effect, always seeking a mate to father more young. Below are three examples, in addition to nine mentioned elsewhere in other contexts in the book: ruffed grouse (p. 141), Anna's hummingbird (p. 69), broad-tailed hummingbird (p. 22), American woodcock (p. 130), sedge wren (e.g., p. 91), marsh wren (e.g., p. 60), bobolink (p. 68), red-winged blackbird (e.g., p. 117), and great-tailed grackle (e.g., p. 166).

Wild turkey

The gobbler "gobbles," much like his distant relative the rooster "crows" (p. 137). The gobble is guttural and chortling and gurgling all at once, descending in pitch, as he tries to convince the females in the flock that he should be the one to father their young (♩86). In this selection, several males gobble from their night roost before departing for the day; in the distance (at 0:28) a second flock also gobbles from its overnight roost.

Dickcissel

Dick-dick-ciss-ciss-ciss-ciss the male dickcissel sings, throughout the day from dawn to dusk. If he and his territory are worthy, he might attract up to six females to nest with him (♩87). With a roughly 50:50 sex ratio, that leaves a lot of less worthy males unpaired, many of whom are yearlings.

More about the dickcissel: "Song (and Call) Dialects" (p. 63).

Yellow-headed blackbird

The yellow-headed blackbird offers some of the strangest guttural and whining sounds heard among birds. Songs are delivered in a jerky, irregular rhythm with strangled contortions of the entire body, sounding much like a "wail of despairing agony." Each male has two different songs. With one, used in nearby territorial disputes, his head is pointed up and to the left (never the right), and with wings slightly spread, he wails out a *kuk—koh-koh-waaaaaaaaaaaaaaaaaaaa,* ending in a long nasal trill (heard at 0:18 in ♩88). With the second song, he responds to more distant males, uttering (e.g., at 0:03) a relatively simple *kak-kak-ka-kaaow,* spreading his wings in a V over his back and exposing the white in them. With these theatrics, a stud male can attract up to eight females to nest on his territory.

There is something special about these blackbirds and their strange songs and unusual displays. Listen to this half hour of hungry fledglings, calling males and females, and of course singing males, and I expect you will become as fascinated with them as I am (♩89).

> ### Explore 15: Singing for more and more mates.
>
> Spend some time quantifying the efforts of an overtly polygynous songbird and you soon realize that a lot of his day is spent singing and addressing females. A dickcissel seems to sing all day long. A marsh wren sings and builds multiple nests for prospective mates throughout the day, and sometimes sings through the night. The singing season of polygynous birds can also extend much later into the breeding season than it does for typically monogamous species, such as a brown thrasher or black-throated blue warbler, in which one male and one female pair.

COURTSHIP SONGS

With much of the birdsong we hear throughout the day, the male appears to be singing on automatic pilot. One song methodically follows another, the male routinely delivering his baseline effort. But when something significant happens, he ups the ante, now singing more rapidly, with more animation, in some way showing his heightened emotions. Such is the case when he courts a female (or tries to impress during the dawn chorus: see "Energized Dawn Singing," p. 112).

House finch

The routine songs of a male house finch are a bright, rollicking, cheery warble, a bit hoarse, with a fast tempo throughout, usually descending on a gentle scale; he occasionally punctuates his song with his signature note, a longer, raspy, down-slurred *veeeeer* (♫90). You can hear these same features in more animated daytime singing (♫91) and in excited singing during the dawn chorus (♫92). But catch him courting a nearby female, his wings now drooped and fluttering as he hops rapidly about. He sings, but sounds far more excited as he alternates songs with high-pitched courtship calls (♫93).

Cassin's finch

Songs of the Cassin's finch are also rich and warbly (♫94), similar to those of the closely related house finch above. But then on occasion he launches into a different kind of song, punctuated with a high, thin whistled note up around 9000 Hz, well above the rich phrases down below, reminiscent of the style of the house finch's courtship song. Hear him now when he appears to be courting a female, with the normal song phrases muted, the high whistles loud and intense (♫95).

Listeners with a practiced ear will also hear imitations of other species in these songs, three of the most noticeable to my ears being notes of a western tanager (heard in ♫602), northern flicker (♫335), and ruby-crowned kinglet (♫387).

Lark sparrow

The lark sparrow has one of the finest sparrow songs in North America, with alternating clear notes and gurgling notes and trills, sounding sometimes like a canary, sometimes like an exalted song sparrow, always rich and thrilling, always "full of life and animation . . . poured forth with great fervor" (♫96).

Listen to him gush while courting a nearby female in early spring, his continuous song now more subdued, as if whispered into her ear—the words "tender" and "affectionate" inevitably come to mind (♫97).

Rose-breasted grosbeak

His everyday songs are spectacular, a rippling series of rich, melodious notes, given with seemingly great energy and enthusiasm (♫98). But then hear him pull out all the stops while courting a female, the song now reaching higher frequencies, with deliciously stretched-out whistles gliding down the scale, a seemingly endless serenade of some of the finest birdsong imaginable (♫99).

More about rose-breasted grosbeaks: "Female Song and Duets" (p. 15).

> ### Explore 16: Listening for courtship songs.
>
> (See website for details.)

SINGING IN THE BRAIN

The brains of true songbirds contain several unique clusters of neurons called song control centers (see figure on p. 33). Two identified brain pathways allow songbirds to learn and produce songs. One pathway (called the anterior forebrain pathway, illustrated here with black lines) enables fine listening and processing, and also guides song learning; signals from the HVC (high vocal center) are transmitted to "Area X," the DLM (dorsolateral medial nucleus), and the LMAN (lateral magnocellular nucleus of the neostriatum). To produce a song, signals (represented by red lines) from the HVC are transmitted to a nerve center called RA (robust nucleus of the arcopallium), which ultimately controls the nerves (12th cranial nerve, nXIIts) that guide the syrinx to produce the intended song.

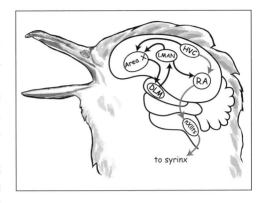

These neuronal pathways enable the songbird to learn its songs, much as we learn to speak (see "Learned Songs of Songbirds, and Babbling," p. 49), and to master complex songs and large repertoires (see "Song Complexity," p. 79, and "Small to Large Repertoires," p. 83). How these song control centers differ among songbirds helps us understand how the brain controls song.

Songbirds with special song learning centers

California thrasher

The larger the repertoire of unique songs that a male songbird can sing, the larger the song control centers in his brain. A brown thrasher, for example, sings well over a thousand different songs (see p. 92) and devotes huge areas in his brain to controlling all that song. So I was excited during 2017 to visit the California thrasher, a close relative of the brown thrasher. In this exuberant and exquisite songster of chaparral communities in California, a mimic as well, I had expected enormous song variety and would imagine a correspondingly magnificent brain

dedicated to controlling it all. Besides, I had read that he typically delivers a song phrase one or two times before hustling on to the next, just like the brown thrasher.

Oh, how the California thrasher surprised me! Each morning, for 15 to 20 minutes during the dawn chorus, the male I chose to study focused on *just one song theme,* with a different theme emerging each morning (by "theme" I simply mean a distinctive and easily recognizable set of sounds that he incorporates into his songs). The themes are best compared to each other in relatively brief excerpts from three mornings (♪100, ♪101, ♪102), but I love the full dawn sequence on each morning, to hear his first song, followed often by a long pause, then scattered songs over several minutes, and eventually some focused singing before he departs his awaking bush for the day (♪103, ♪104, ♪105). Wow! I know of no other songbird who, with apparently so much to say, restricts himself to such a modest performance on a given morning. How many mornings would I need to listen to this thrasher before hearing him return to something he had already sung? I don't know. Nobody knows.

In disbelief, I sought out other California thrashers, but heard the same pattern from them as well (e.g., ♪106).

Explore 17: The puzzling case of the California thrasher.

It would be fascinating to find where a California thrasher awakes, morning after morning, and listen to (and record) all that he can do. Would he ever return to something he had sung before? If so, how many days of listening would it take to hear some repetition? If never, does he compose original songs each day, settling on some theme that most satisfies him, or has he a seemingly infinite store of set themes to choose from? Thrashers—what extraordinary singers!

Marsh wren

Western marsh wrens from California learn about 150 different songs (♪107), eastern males from New York only about 50 (♪108). Accordingly, the song control centers in western birds are 30 to 40 percent larger than those of eastern birds, even though the western and eastern birds have about the same body size. Also, the more songs an individual

  male learns, the more complex his neurons become in his song control centers.

More about the marsh wren: "Song (and Call) Matching" (p. 58) and Explore 28 (p. 61).

Carolina wren

The male Carolina wren learns 30 to 40 different songs and has large song control centers in his brain; the female, with only an innate, non-learned, rattling call (p. 17), has no song and no identifiable song centers. A male normally repeats one of his songs many times before introducing another, but

listen to this excerpt of his efforts, with just one example of each of eight different songs (♫109). The mnemonic *TEA-KETTLE* is often used to describe the individual phrases in the song.

Among some tropical wrens who are close relatives of the Carolina wren, both males and females sing, and, as one might expect, the female has song control centers nearly as large as those of her mate. If Carolina wrens duetted like some of their tropical cousins, the male and female would coordinate their singing precisely, singing as if they were one bird. Imagine taking a male Carolina wren *TEA-KETTLE* song and bisecting it, assigning the *TEA* to the male, the *KETTLE* to the female. In the simulated tropical-type duet I have prepared in ♫110, I have placed the *TEA* on the left track, the *KETTLE* on the right (listen with headphones), as if you were standing between the two birds, the male to the left, the female to the right.

More about the Carolina wren: "Female Song and Duets" (p. 15), "Learned Songs of Songbirds, and Babbling" (p. 49), and "Mimicry by Mockingbirds" (p. 75).

Brains of non-songbirds lack song learning centers

The song control centers that are found in all songbird brains are absent in the brains of birds that do not learn their songs, such as the flycatchers ("suboscines"), who are the closest relatives of the songbirds.

Eastern phoebe

The eastern phoebe sings its name in two different song forms, the *FEE-bee* with the buzzy *bee* and the *FEE-b-bre-be* with the stuttered *b-bre-be* (alternated in ♫111). Unlike songbirds in the same passerine order, this typical suboscine flycatcher does not learn its songs and has no identifiable song control centers in the brain. The alder flycatcher and willow flycatcher likewise do not learn their songs (p. 45 and do not have song control centers.

More about phoebes: "Inborn Songs" (p. 44); compared to two other phoebes in the genus *Sayornis* (pages 84–85).

California quail

Although we might say that "chickens" (p. 137), including quail, grouse (p. 141), prairie-chickens, turkeys (p. 29), and their relatives, can "sing," they lack the song-learning centers in the brain, and it is believed that their drums, crows, gobbles, and other display sounds are all inborn. Unlike the songbirds, individuals of these species do not acquire their sounds by imitating other adults of their species. This California quail, for example, delivers an innate

CAAAH . . . CAAAH . . . chi-CAH, chi-CAH, chi-CAH-go, chi-CAH-go (♫112).

Western gull

Gulls show no evidence of learning their calls, nor of the songbird type of song control center (♫113).

Other groups that do not learn their sounds and have no song control centers are owls (pages 124–125), swifts (p. 13), and doves (p. 45), all in nonpasserine orders.

Explore 18: Bird brains and special song control neurons.

(See website for details.)

Sage thrasher

Learning, memory, and recall are truly remarkable abilities for many songbirds. Especially astonishing was this sage thrasher who was into the eighth hour of his singing and my listening when he produced two snippets of song sparrow imitations that he had not used before. No song sparrows were anywhere near, so this imitated song was dredged up from some distant memory that had been stored in the thrasher's brain. Listen first to the song of a song sparrow and then to the two sage thrasher snippets, arranged to help you identify the mimicry in the thrasher song (♫114). The thrasher's rapid-fire song program requires that each snippet be about a half second, so to retain his cadence he splits the more leisurely sparrow song in two, then rushes through each half. Test your ears further by listening for those two snippets in the intact 20-second song of the thrasher (♫115).

More about the sage thrasher: "Song Complexity" (p. 79) and "Night Singing" (p. 123).

> ### Explore 19: Memories of birds and human listeners.
>
> Test both your memory *and* that of a singing bird. Find a thrasher (or a relative, such as a catbird or mockingbird) who sings endlessly, and listen to him until you hear a sound so distinct that you will be sure to recognize it when he sings it again. Then begin your Big Listen. Start your stopwatch, or count the number of songs until you hear your chosen song again, or do both. How long did it take him to return to your chosen song? How many other songs did he sing in the meantime? Or, in case you can't get out to find your own singing bird, use the long recordings I provide for a number of species, such as the red-eyed vireo (p. 98), gray catbird (p. 49), brown thrasher (p. 92), northern mockingbird (pages 73–75), sage thrasher (Explore 61, p. 127), and western meadowlark (pages 150–151).

NOT ONE BUT TWO VOICE BOXES

Our voice box (the larynx) sits atop the trachea (windpipe) within our throats, but the avian syrinx lies at the base of the trachea, within the chest, with one half of the syrinx atop each bronchus as it leads to the left or right lung. Each half of the syrinx is controlled independently by a nerve that runs from the brain's song control centers down the side of the neck. Singing, then, is really precision breathing, as dictated by the song control centers.

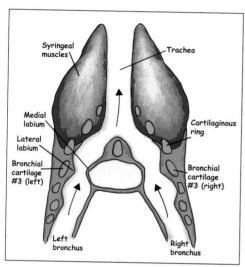

This figure shows a highly simplified, cross-sectional schematic of a typical songbird syrinx. When a bird exhales during normal breathing, air flows freely through the bronchi and out the trachea (arrows). When the bird chooses to produce sound, special, elongated cartilaginous rings (bronchial cartilage #3) rotate into the airstream to produce the desired effect. If the sound is to come only from the left side, the right #3 bronchial ring rotates to completely close off the air passageway on the right; the left #3 bronchial ring simultaneously rotates, but only partway, narrowing the gap between the soft tissues on the outside (lateral labium) and inside (medial labium). These soft tissues then vibrate in accordance with the nerve impulses directed by the brain.

It boggles the mind to realize all that has to happen in this syrinx at the pace required of, say, a wood thrush song, in which the two voices are engaged simultaneously (see next section)! The singer can also alternate between the two halves of the syrinx, using them in rapid succession. The syringeal muscles are, in fact, "superfast," among the fastest twitching muscles known among animals. Curiously, for unknown reasons, birds seem to use the left side of the syrinx for the lower frequency sounds, the right for the higher frequencies.

Wood thrush

The song of this eastern forest maestro is heard in three parts: (1) a couple of soft, introductory *bup bup* notes; (2) a "prelude" of low, fast-paced, pure tone notes; and (3) a "flourish" of typically higher, trilled sounds, often sounding noisy or percussive to our ears. Surprisingly, it is in that flourish where the wood thrush excels, using his two voices *simultaneously* to produce extraordinary harmonies that we cannot appreciate with our unaided ear (see "What Birds Hear," p. 41).

Listen first to a normal sequence of wood thrush song (♫116), and then imagine yourself a wood thrush as you hear the details in the first six songs when they are slowed to one-quarter and then to one-eighth their normal delivery speed (♫117, ♫118). Next, let's separate the two voices in five different flourishes, so that in our left ear we hear the lower voice from the left side of the syrinx and in our right ear the higher voice of the right, all at one-eighth normal speed so our ears don't miss a beat (♫119). (Use headphones for best effect.)

More about wood thrushes: "How a Repertoire Is Delivered" (p. 94) and "How Birds Go to Roost and Awake" (p. 108).

Varied thrush

The varied thrush's songs are eerie and penetrating; some of us interpret them as melancholy, others as thrilling, but always mysterious. Each song consists of two simultaneous voices that we can almost imitate by whistling and humming at the same time, maintaining a constant pitch for a second and a half. Go ahead and try it! And then he sings another song, and another, maybe half a dozen distinct songs before he starts repeating himself, all arranged in a sequence that emphasizes the striking contrast from one song to the next (♫120).

More about varied thrushes: "The Music in Birdsong—Dissonance" (p. 159).

Northern cardinal

When a cardinal slurs what sounds like a single whistle through a broad range of frequencies, either up or down, he begins with one side of the syrinx and seamlessly transitions to the other half to complete his slur. The separate notes of the two voices are stitched together by precision breathing through the two syringeal halves at about 3500 Hz, which roughly corresponds to the highest A note on a piano. The two voices can also be used to rapidly alternate two notes at different frequencies, or to deliver strings of high and low notes in a more leisurely sequence.

Listen to three examples for each of four different songs excerpted from a long performance by a male cardinal (12 total songs, ♫121), then to just one example of each of those four different songs (♫122). Next, we slow each of those four examples down to half (♫123) and then to one-quarter (♫124) the normal speed. Listen carefully, and at these slower speeds you can increasingly hear a minor discontinuity in the slurred whistles that introduce the songs (especially from 0:00 to 0:10) as the male transitions from his right to his left voice. If you upload these songs into Raven Lite, you can see that slight break where he transitions from one voice to the other (e.g., look at 0:04.8 in ♫124). Throughout these songs, most of the low notes are likely produced by the left side of the syrinx, the high notes by the right, though knowing for sure would require careful internal study of how the cardinal breathes through the syrinx from the left and right lungs.

More about cardinals: "Female Song and Duets" (p. 15), "Song (and Call) Matching" (p. 58), "Small to Large Repertoires" (p. 83), and "How Birds Go to Roost and Awake" (p. 108).

Brown-headed cowbird

The cowbird is a remarkable singer, though our unaided ears fail to appreciate the vocal gymnastics involved. He begins by alternating his left and right voices with notes so soft and so low that our ears strain to hear them (lower than middle A on the piano, down to about 300 Hz), then in a hundredth of a second leaps with his right voice to an octave above the highest note on the piano, with notes so high they once again challenge our hearing. With headphones, listen first to normal songs from a flock of seven male cowbirds perched on a wire five yards overhead (♪125). Then listen to three songs excerpted from that sequence (♪126), first at normal speed, and then at quarter speed, which enables our ears to detect many of the exquisite details we had missed before.

Can't get enough of cowbird magnificence? Here's a flock of dozens in spring song (♪127); I love listening to an excerpt of that recording slowed to quarter speed (♪128). You can import these songs into Raven Lite and play with them some more (see "How to Hear and *See* Birdsong," p. 178).

More about cowbirds: "Song (and Call) Dialects" (p. 63).

> **Explore 20: Seeing the two voices of birds in sonagrams.**
>
> (See website for details.)

WHAT BIRDS HEAR

Extensive testing of different bird species in the laboratory has revealed that most of them hear roughly the same range of frequencies that we do, with some exceptions (see two examples below). Where the songbirds excel, the laboratory experiments show, is in their "temporal resolving power": They hear details of complex sounds while we hear only a blur. When we slow the songs down to about one-quarter speed,  or even slower, we begin to appreciate what the birds can hear. For that reason, we have already been slowing many songs to get to know them better.

Henslow's sparrow

His song is a half-second hiccup, or *tsilick* (♪130), "one of the poorest vocal efforts of any bird" thought Roger Tory Peterson, who, like all of us humans, had a clumsy ear for appreciating this song's dips and jogs and feints and abrupt turns in 1/100th of a second. When one slows this song down four or eight times ($\musicalnote$131) to reveal for our ears the details that the birds themselves must hear, we are astonished to hear a masterpiece of sliding whistles, beginning high and gliding down the scale.

Pacific wren

An eight-second series of high-pitched, tiny notes and tinkling trills ($\musicalnote$132) from this minuscule songbird passes quickly, sounding shrill and almost percussive, but the glorious details are lost on our ears. Slow the song down four times and we can better appreciate him; slow it eight times and our ears are even more grateful ($\musicalnote$133). A young bird learns the minutiae of these songs from singing adults, copying the songs bit by tiny bit, perhaps the best proof of all that these wrens can hear all the details.

Rock pigeon (common pigeon)

A special domesticated line of this common pigeon, the homing pigeon, has been shown to hear frequencies far below our range of hearing—down to about one-tenth of a cycle per second, while the human's lower limit is often given as 16 to 20 cycles per second (also expressed as 16 to 20 Hz). These low-frequency sounds, which carry enormous distances, may be used by pigeons to generate an auditory map of their known world and then use that map to find their way home. Listen here to sounds of a male arriving at the nest and being greeted by his mate; these vocalizations are all within the range of human hearing, but just try to imagine the pigeon's auditory world and what else it hears, well below our capability ($\musicalnote$134).

Red junglefowl (barnyard chicken)

Another surprise: The common barnyard chicken can also hear especially low frequencies, much like the pigeon. How an egg-laying hen (♫135) might use this ability is unknown, but consider a distant relative of hers, the ruffed grouse (p. 141). We hear the grouse's drum as muffled, and if close enough we feel it resonate in our chest better than we hear it with our ears, because most of the energy in the

drumming lies far below our range of hearing. It is a good guess that the grouse, like the domesticated chicken, can also hear well into those low frequencies, thus enabling each grouse, male or female, to monitor the drumming activity of all males within grouse earshot, over distances our ears cannot fathom.

More about the barnyard chicken: "Each Individual Has Its Own Song" (p. 137).

> ## Explore 21: Playing songs at slower speeds, lower frequencies.
>
> Listening to birdsong slowed four times, down to one-quarter speed, probably brings us as close as we can come to appreciating the intricacies of birdsong as the birds themselves hear it. This book's website offers hundreds of sounds that you can download and import into the software program Raven Lite (see p. 178), where you can slow the songs to whatever speed you wish as you watch the sonagrams dance across your computer monitor. I suggest beginning at one-quarter speed (0.25), and then go from there. Perhaps start with the thrushes, such as the Townsend's solitaire, eastern bluebird, varied thrush, veery, Swainson's thrush, wood thrush, and hermit thrush. You will soon realize all the more why they are considered such special songsters. As you listen and watch the sonagrams, can you find clear examples of two strikingly different sounds being sung simultaneously, from the two voice boxes?

4. How a Bird Gets Its Song

INBORN SONGS

In some species, songs are inborn, or innate. It is as if the songs of the adult are somehow encoded in the DNA, and the developing embryo in the egg already has the instructions it needs to sing the proper song. Because all birds within these species have essentially the same song genes, the songs vary little from one individual to the next and from one location to another, even across vast stretches of the continent. If songs of these species with innate songs are found to be different in a particular geographic area, especially an isolated area, professional ornithologists are likely to take note, wondering if those vocally different birds aren't actually a different species with not only different song genes but different overall genes (as happened with eastern and Mexican whip-poor-wills; see page 27).

The first three examples that follow are from nonpasserine orders, in which most sounds are believed to be innate. The last three examples are suboscine flycatchers in the order Passeriformes. Songs of flycatchers and other suboscines (except bellbirds in the genus *Procnias*) are believed to be inborn, not learned like those of songbirds.

Northern bobwhite

Without song-learning neurons in the brain, this tiny quail-chicken delivers his familiar name throughout the East, *oh BOB WHITE!*, beginning softly and ending sharply on the *WHITE!* It is essentially the same song everywhere these quail are heard

(♫136). You can explore the internet and find other examples of the bobwhite's song, but they'll all sound like the classic *oh BOB WHITE!*

Mourning dove

This dove is named for its mournful, owl-like song, *cooowaah, cooo, coo, coo* (♫137). As you listen, realize that the instructions for the *cooowaah, cooo, coo, coo* are embedded deep in the genes. On the rare occasion when a male and female of two different dove species mate, the mixed-up genes of their offspring produce perfectly mixed-up songs, exactly what one would expect when half of the song genes are from each parent.

In early April in northern Michigan, the dawn chorus is fairly quiet, except for the mourning doves; in ♫138, one sings in earnest for seven minutes, with other mourning doves flying about.

More about the mourning dove: "Mechanical (Nonvocal) Sounds" (p. 21).

Eurasian collared dove

The Eurasian collared dove also has an innate song (♫139) and, like the mourning dove, also uses the mechanical sounds of whistling wings on takeoff (at 3:12). Arriving in Florida during the 1980s, this immigrant has extended its range all the way to the Pacific Ocean.

Alder flycatcher, willow flycatcher

These two flycatchers look nearly identical, and it was only in the 1970s when ornithologists realized that the bird then called the "Traill's flycatcher" actually consisted of two unique groups, each with its own song. In one group (now named the alder flycatcher) a male sings a single song, something like *free-BREER-a,* while in the other group (willow flycatcher) he sings three different songs: *FITZ-bew*

(with a sharp first syllable in the two-parted song), *FIZZ-bew* (fizzy first syllable), and *creet* (simple rising note). Later studies showed that the songs of these flycatchers were inborn, or hard-wired, so that the songs reliably reflect the genetic makeup of the singer.

For the alder flycatcher, listen to males from Alaska (♫140), Massachusetts (♫141), and Maine (♫142), and for an easier comparison, listen to one song from each of the three birds at these distant locations (♫143). The same song genes throughout the range guarantee the same song everywhere.

For the willow flycatcher, listen to birds from Missouri (♫144), Kentucky (♫145), and Massachusetts (♫146), all with the same three songs. And if these brief selections aren't enough for you, try the nearly hour-long recording of a male through his entire dawn chorus (♫147). In these recordings, it is the same songs, the same song genes, and the same species over eastern North America. "Wait!" says the astute bird lover. "What about birds from the West, beyond Missouri?" Good question. Slightly different songs in the West mean slightly different genes, with some ornithologists lobbying to split eastern and western willow flycatchers into two or even three species (see page 142).

More about alder and willow flycatchers: "Each Species Has Its Own Song" (p. 140) and "Song Changes over Evolutionary Time, from Species to Species" (p. 147).

Eastern phoebe

Every eastern phoebe knows two songs: a *FEE-bee* with a raspy *bee* and a *FEE-b-bre-be* with a stuttered *b-bre-be*. Because the phoebe's song genes are essentially the same everywhere, the songs of this common backyard flycatcher vary little from one bird to the next, from Missouri (♫148) to Massachusetts (♫149, ♫150) to Virginia (♫151) to Texas (♫152). For a ready comparison, in ♫153 I provide one *FEE-bee* and one *FEE-b-bre-be* from each of the five locations.

During a fast-paced dawn chorus, the male often alternates the two song forms; as the pace slows, a string of the *FEE-bee* songs is typically followed by one *FEE-b-bre-be*. And just when I thought I knew phoebes, I encountered a bird who sang a dawn chorus with hundreds of stuttered *FEE-b-bre-be* songs and not a single *FEE-bee* (♪154). What was that all about?

More about phoebes: "Singing in the Brain" (p. 32); compared to two other phoebes in the genus *Sayornis* (p. 84).

> ### Explore 22: No dialects in the innate songs of flycatchers, doves, etc.
>
> Appreciate for yourself how songs of these birds remain relatively constant over much of the North American continent. For quail, doves, and flycatchers, listen to songs available on the internet and just try to find outlying geographic areas with distinctive songs. You won't find them, as you are hearing the power of the genes to guide song development into a very narrow outcome. Or if you do, chances are the professional ornithologists are aware of those differences and are contemplating whether the song differences are sufficient to split the outlying birds into a different species.

IMPROVISED SONGS

Songbirds do not have innate songs like those of their suboscine relatives, but a few songbird species do enter the world with something like genetic instructions on how to improvise proper songs (instead of how to imitate them precisely—see next topic). The young bird innately knows much of what his songs should sound like, and, probably combined with some experience listening to adults, he uses this information to compose his own songs. As a result, no two individuals have exactly the same songs, and this kind of creativity can lead to impressively large song repertoires. One could say that the young songster is learning from himself, as he listens to his own songs and revises them accordingly, with all those song control centers in his brain fully engaged in the process.

Red-eyed vireo

The red-eyed vireo seems to sing his name, *vireo,* with an infinite number of inflections and enunciations. Listen carefully to an individual, however, and you will hear him return every 20 or so songs to a recognizably unique song or two that stand out from the others (see p. 98). Other vireos in the neighborhood won't have that same song, because that odd song, and probably most of his more standard *vireo* songs, was created in the improvising mind of the vireo.

You can practice finding unique songs among these three sample recordings: ♫155—try the burry, descending song at 0:12; ♫156—this bird is tough, because no songs stand out as especially different; ♫157—try that cute whistled note at 0:03 (for a longer session of that bird, listen to ♫158).

More about red-eyed vireos: "Learned Songs of Songbirds, and Babbling" (p. 49) and "How a Repertoire Is Delivered" (p. 94).

> **Explore 23: Improvised songs of red-eyed vireos.**
>
> (See website for details.)

Sedge wren

Cut-cut-cuta-trrrrrrrrrrrrrrrrr (♫159). With the same dry chatter introducing a slow-to-fast trill of repeated phrases, all sedge wren songs are of the same pattern, all declaring "I am a sedge wren." But use computer analysis (e.g., Raven Lite) to compare the 100 or more different songs in the repertoire of one male with those of his neighbor, and you will discover, as I did, that the two birds have entirely different songs. The songs among different males are like snowflakes, all with the same overall pattern but with no two exactly alike, because each individual improvises his own unique repertoire.

How do I know that? Dumbfounded at how male sedge wrens shared no songs with their neighbors while their close relatives the marsh wrens shared almost all of their songs (p. 60), I raised some baby wrens of both species in the laboratory. The baby marsh wrens imitated perfectly the marsh wren songs that I played to them over loudspeakers, but the sedge wrens imitated none of the sedge wren songs they heard, instead making up a healthy repertoire of their own.

Perhaps he improvises his songs because sedge wrens in North America are somewhat nomadic, never reliably breeding in the same place from year to year, so there is no incentive for a male to invest a lot of energy in the precise imitation of males at one location. (Compare this developmental strategy with that of its close relative, the non-nomadic, song-imitating marsh wren, p. 60.)

More about sedge wrens: "Small to Large Repertoires" (p. 83), "Each Species Has Its Own Song" (p. 140), and "The Music in Birdsong—Improvisation" (p. 159).

Gray catbird

The catbird is also a master improviser. A young catbird who is raised away from all adult singing catbirds improvises an entirely normal set of his own songs, up to 400 of them! Listen to this catbird race through his songs and appreciate the creative mind at work (♫160). Go back and listen again to the

brown thrasher (♫84, p. 28), a distant relative in the same songbird family (see "Who's Who?," p. 176), and realize that he, too, can create his own large, unique repertoire by improvising on his own. Gray catbirds and other "mimic-thrushes," such as the thrashers, are versatile in that they not only improvise songs, but on occasion also mimic the songs of other species (p. 92). Listen to this catbird (♫161) in the depths of Hells Canyon, on the border of Oregon and Idaho, return repeatedly to sing the *FITZ-bew* of a willow flycatcher (listen again to the flycatcher in ♫144).

More about catbirds: "Birds Sing and Call" (p. 6).

LEARNED SONGS OF SONGBIRDS, AND BABBLING

All songbirds probably improvise to some extent, but they are better known for their ability to learn. Young birds imitate rather precisely the songs of adults, just as we humans imitate the language of our parents and other adults. Without that influence from accomplished adults, both these songbirds and we humans would have nothing intelligible to say. Three songbird examples of species known to imitate are included here: the Carolina wren, the song sparrow (especially western, nonmigratory birds), and the indigo bunting. This kind of imitative learning occurs not only in songbirds, but also in parrots and some hummingbirds, and also in the bellbirds in the suboscine genus *Procnias.*

All songbirds (including those who improvise) practice their songs for extended periods, beginning when just a few weeks old and often completing their task nearly a year later, during the following spring. When practicing, a young bird listens carefully to himself, trying to match his voice to some combination of (1) the adult's voice that he has memorized and stored in his brain's song control centers and (2) the nature of the song that he innately knows. Hearing is essential, as neither a deaf human nor a deaf songbird can ever learn to speak or sing accurately. What fun to listen to the rambling babbling (or "subsong," as it is called) from both the imitating and the improvising songbirds here and imagine all that is going through the youngster's mind as he tries to get it right. (Ditto for a human baby, of course!)

Babbling during late summer and early autumn

Warbling vireo

From the treetops during spring, the adult warbling vireo delivers a spirited, undulating volley of jumbled notes (♫162), rising and falling so fast that it seems the singer trips over himself, the adult having mastered his particular version of *If I SEES you, I will SEIZE you, and I'll SQUEEZE you till you SQUIRT!* An August youngster, about two months old, lacks the finesse of the adult but is immediately recognizable as a warbling vireo in training (♫163).

More about warbling vireos: "Each Species Has Its Own Song" (p. 140).

Red-eyed vireo

The adult red-eyed vireo is a master improviser (see page 48); imagine the journey that this adult male (♫164) has taken to become the accomplished songster that he is. Perhaps it was only the previous year that he had hatched, and within a few weeks had left mom and dad and was already scratching out feeble attempts at song. By August (♫165), when about two months old, his ramblings were improving and already recognizable as those of a red-eyed vireo.

More about red-eyed vireos: "Improvised Songs" (p. 47) and "How a Repertoire Is Delivered" (p. 94).

Carolina wren

The adult male sings one of his *VICTORY!* or *TEA-KETTLE!* songs many times before switching to another (see page 35; another two examples here, ♫166, ♫167). Compare the performance of adults to that of a two-month-old male during a midmorning practice session (♫168); this youngster stumbles from one bumbling song to the next, no two alike. Listen also for his raspy, down-slurred *jeeeer* call, the details of which may also be learned, as it can vary from one distant location to another.

More about the Carolina wren: "Female Song and Duets" (p. 15) and "Singing in the Brain" (p. 32).

Song sparrow

Songs of the song sparrow are heard as lively and cheerful, beginning with a few sharply enunciated notes, followed by several contrasting phrases. An adult male learns about eight different songs, and, like the Carolina wren, delivers each one many times before switching to another (settle back and listen to this half-hour romp in ♫169). The singing attempts of this four-month-old during October are a jumble of uncertainty, with successive attempts at the same sound still highly variable, his babbling often blurting out seemingly random sequences of song bits stored in his brain (♫170).

White-throated sparrow

An adult white-throated sparrow has just one song, delivered as the purest of unwavering whistles, typically beginning with two or three longer whistles and then rising or falling to brief triplets (sometimes doublets), such as *ohhhhh sweeeet Can-a-da Can-a-da Can-a-da* (♪171). During October, in ♪172 a young migrant at first practices his two longer whistles, but is uncertain as to whether the second should be higher or lower than the first; in his last attempt, there is just a hint of the *Can-a-da* triplet.

More about white-throated sparrows: "Song (and Call) Dialects" (p. 63).

Explore 24: The practice singing of young songbirds.

Search for young birds on your own, at the best time of year to find them where you live. In the Northeast, I begin listening carefully during August. I find a birdy place and just stand quietly, perhaps with eyes closed, concentrating, listening for anything out of the ordinary. The first musings of young male birds are oh so soft and amorphous, so easily overlooked, and if heard might not even be identifiable to species. Their efforts gradually improve, however, and by September and October, songs can already be very adultlike. But there is always some irregularity in the songs of these young birds, something that gives them away. Perhaps it's a wavering note, or a sudden break in frequency where there should be none. Or, for a species in which a given song is typically delivered many times, successive songs are slightly to hilariously different, as the song has not yet been mastered.

More practice the following spring

During spring, sometimes one cannot be sure if the unsteady singer is a young bird coming into adult song for the first time, or an older bird resuming his practice for a new season.

Indigo bunting

Over three to five seconds, an adult male delivers his husky, mostly paired phrases, *fire fire where where heeere my my run run run faster faster safe safe pheeeewwww* (♫173). During early spring, a young male, recognizable by his mottled blue and indigo plumage, returns from migration and rambles on incessantly (♫174, ♫175). He typically practices far more song phrases than he can possibly use, and is still missing some that he will need to learn from his adult neighbors in order to match their songs.

More about indigo buntings: "Big Decisions: When, Where, and from Whom to Learn" (p. 56) and "The Music in Birdsong—Diminuendo" (p. 157).

Tennessee warbler

Migrant warblers offer good opportunities to hear spring males practicing their songs. An adult Tennessee warbler, for example, has one loud, staccato song, repeated over and over, typically a three-parted (sometimes two-parted) *ticka ticka ticka, swit swit, sit-sit-sit-sit-sit-sit,* becoming louder and faster from one part to the next (♫176). This spring migrant in Western Kentucky (♫177), either a young bird or

an adult reviewing his song from the year before, lacks the consistency that will soon be achieved by the time he arrives on his breeding territory far to the north.

American redstart

High-pitched, thin and sharp, and what a wicked variety of songs these redstarts have. An adult male has several different songs that he uses during the intense dawn chorus, and then an extra song that he reserves for more routine singing during the day, probably more to communicate with females (see also p. 139). In this example (♫178), hear the five

different songs this male uses during the dawn chorus (up to 3:00), and then five renditions of his daytime song (after 3:00); each of these song variants is repeated precisely from one time to the next. How uncertain is this May migrant (♫179), his yellowish feathers identifying him as a yearling; no two efforts sound alike, but eventually he'll master the songs he has chosen.

More about the redstart: "Each Individual Has Its Own Song" (p. 137).

Magnolia warbler

Magnolias are known for their short, pure, loud songs, such as *weeta-weeta-weetsee,* often only three phrases, the third rising higher at least initially. Each adult male has his own version, but just one, which he sings with unerring regularity (♫180). This May migrant (♫181) seems to have kept all options open, as there is as yet no consistent hint as to the exact form his final song will take. I bet he is a yearling, as I'd think that an adult at this time of year would know better what his song is to be.

White-throated sparrow

One can never quite be sure what a white-throated sparrow is going to come up with. Many of them (in the East, anyway—see page 67) have the *Can-a-da* triplet, but others seem to have a mind of their own, or possibly they are in tune with small dialects where different versions of the white-throat song are standard. From northern Maine, here's an odd-sounding adult who doesn't quite fit what we think of as the norm (♫182). Listening to white-throats (of unknown age) practice in the spring is a wonderful pastime; enjoy this aspiring musician from Virginia (♫183).

More about white-throated sparrows: "Song (and Call) Dialects" (p. 63).

Carolina wren

The greatest number of times I have heard an adult male Carolina wren repeat one of his song types before moving on to another was almost 500; during a late-summer morning, that performance took more than an hour. This adult male sings nearly 100 renditions of the same song in about seven minutes (♫184). Then here's a puzzle: During early June in southern Illinois this male was bumbling along in plastic song (♫185). He must have been from an April hatch, I reasoned, and therefore only about two months old, because any bird hatched the previous year would have perfected his adult songs well before early June.

More about the Carolina wren: "Female Song and Duets" (p. 15) and "Singing in the Brain" (p. 32).

> ### Explore 25: Listening for practice singing during spring.
> (See website for details.)

5. More about Song Learning

BIG DECISIONS: WHEN, WHERE, AND FROM WHOM TO LEARN

As early as two weeks of age, when just leaving the nest, many young songbirds begin to listen intently and memorize the adult songs that they hear. The most intense memorizing seems to occur during the first two months of life, but many yearling males further refine their songs the following spring when they settle onto their first breeding territory.

Although learning may begin while the young bird is still with the singing father, in many species dad's songs are eventually rejected when the young male settles to breed some distance away and opts for a song that is more like that of his immediate neighbors. It is important that he sings the local dialect and fits into his own unique singing neighborhood, so that he has the same songs as his singing neighbors. And just why is that important? Hmmmm . . . wish I knew! It is likely important for the countersinging they do, being able to match each other's songs (p. 58). Maybe females want assurances that a male is not a vagrant, but instead is committed to her particular singing neighborhood. The fact that males of so many species do it verifies its importance, but the exact reasons why are unknown.

Chipping sparrow

Each male chipping sparrow has just one song of identical, repeated phrases, but what a great variety of songs among birds, ranging from dry, chipping rattles to musical sweeps of pure tones up or down the scale. Listen closely to a population of chipping sparrows and, among all this variety, you will often hear two neighboring males with identical songs, the result of one young bird having learned rather precisely the one song of his adult neighbor.

Listen to the fine variety of songs among these four pairs of chipping sparrows, each pair belonging to its own miniature song dialect of just two or three birds. These four initial recordings have only one song from each of the two birds (♫188, ♫189, ♫190, ♫191), but if you relish longer listening sessions for each of the males, eight longer listens await you as well (♫192 through ♫199).

More about chipping sparrows: "Energized Dawn Singing" (p. 112).

Indigo bunting, lazuli bunting

A young indigo or lazuli bunting also learns the song of his adult neighbor. The birds migrate, but they typically return to breed on the same territory throughout life, and as a result a small cluster of males can come to have nearly identical songs. Here are examples of two neighboring indigo buntings with essentially identical songs from Massachusetts (♫200; one song apiece, as with the chipping sparrows), Virginia (♫201), and South Carolina (♫202). You can also choose longer listens from each of the six males (♫203 through ♫208).

For lazuli buntings, compare the songs of the two neighbors in ♫209. The song of the first male consists of three phrase types; overall, it feels rapid and ends abruptly, as do so many lazuli bunting songs. The song of the second male begins with the same three phrase types, but then he doubles the length of the song by adding several more phrases. It is entirely possible that the first male was capable of singing the longer song too but just chose, for whatever reason, not to do so when I was there.

More about buntings: "Learned Songs of Songbirds, and Babbling" (p. 49) and "The Music in Birdsong—Diminuendo" (p. 157).

Common yellowthroat

The patterns of song development for the chipping sparrows and buntings have been carefully documented by ornithologists. Once one is sensitized to this pattern of only a few close neighbors having identical songs, one can listen more closely for examples in other species. This pattern also occurs among common yellowthroats, as on occasion one finds two neighboring yellowthroats with essentially identical, unique songs, resulting almost certainly from a young male copying the one song of his adult neighbor. Here are four examples of two neighboring yellowthroats with similar songs: eastern Massachusetts (♫210), western Massachusetts (♫211 and ♫212), and Minnesota (♫213). Longer listens are also available for each of the eight individuals (♫214 through ♫221).

More about the yellowthroat: "Birds Sing and Call" (p. 6) and "Song Complexity" (p. 79).

> **Explore 26: Neighboring male songbirds learn from each other.**
>
> (See website for details.)

SONG (AND CALL) MATCHING

When neighbors share the same small or large repertoire of learned songs or calls, they often choose to match each other, with one bird selecting from its repertoire the same song or call that another is using. Why they match each other is not well understood. Maybe they match simply because by doing so they immediately reveal for listeners that they are well established in the community and belong to the local dialect. Or perhaps by matching, competing males on neighboring territories directly address each other, putting each other on notice, so to speak, maybe with stronger intentions or motivation than if they chose not to match. Who matches whom in rapid exchanges, such as among western marsh wrens, might reveal something of the relationship between the two interacting males, with perhaps the leader in the exchanges the dominant individual, the follower the subordinate. Whatever the reason, we often hear this kind of matching if we keep our ears open.

Common raven

The raven is the ultimate songbird, by far the largest, and with a keen intelligence. The variety of sounds made by ravens is legendary, from the twang of a tuning fork to the resonance of a fine bell to a deep, throaty *croak* enunciated in so many different voices. Ravens often match each other's calls, as if acknowledging each other or simply agreeing to agree, but exactly when or why they choose to match or not match each other remains a mystery.

Listen to the minds of the ravens in these two examples. In the first (♪222), one bird calls until 0:16, when a second bird joins, giving the same call; eventually (about 0:52) they both switch to a second call, again matching each other (after that, you are on your own). In the second example (♪223), you will also hear typical call matching at 0:12.

> ### Explore 27: Call matching by ravens, jays, and other corvids.
>
> With a raven or any of his corvid relatives (members of the family Corvidae are *everywhere,* and include jays, scrub-jays, magpies, the nutcracker, and crows—see page 174), focus first on the calls of one individual. Realize that from its vast repertoire of calls, it has chosen one with these particular qualities for the occasion. Then, when you are ready, expand your field of hearing to what other individuals of the species are doing. Do they "agree" or "disagree" with each other? If your bird changes its call, does it do so to match the other birds, or do they match it? What patterns emerge over time?

Northern cardinal, pyrrhuloxia

The cardinal is known for its brilliant, loud, slurred whistles, some rising, some falling, some repeated slowly, some quickly, and with great variety, and they are all packaged into a dozen or so different songs that each male has learned in his small neighborhood (the same could be said for females, p. 19, but the focus here is on male exchanges). Listen to a male cardinal, memorize his song, and then listen for the echo in the background, because more likely than not at least one male on a neighboring territory is responding with the same song from his own repertoire.

In the first example here (♫224), listen to the matching cardinals during an early morning in Virginia, where the cardinal is the state bird (best heard in headphones). Hear how the foreground bird comes to match the background bird from 0:30 to 1:20, and again from 1:52 to 2:40. In the second example (♫225), a male cardinal awakes with sharp *tik* notes, then matches from 1:10 to 1:14 the songs of the male in the distance; the distant male soon changes his tune, and eventually (by 2:40) the foreground male matches him again (until 5:00; check out for yourself what happens the last two minutes).

The cardinal's close relative in the Southwest is the pyrrhuloxia, which also engages in this kind of matched countersinging. Listen to the interactions between the foreground and background birds in ♫226. All the singers are believed to be males, but I wasn't able to see all of them to confirm that.

More about cardinals: "Female Song and Duets" (p. 15), "Not One But Two Voice Boxes" (p. 38), "Small to Large Repertoires" (p. 83), and "How Birds Go to Roost and Awake" (p. 108).

Marsh wren

The fast-paced, complex singing of western marsh wrens offers a challenge for the human ear and mind. Males can learn 150 different songs from each other, and as they race through their large repertoires, they often choose the one song that matches exactly what the neighbor has just sung. This matched countersinging can involve several birds and continue for up to ten exchanges, after which the birds disengage and attend to other singing business.

In this first example (♫227), the males match each other throughout, but to make it easier for you I have excerpted and isolated the eight best matches from that longer session (♫228); sometimes the foreground bird matches the background bird, sometimes vice versa. In the second example seven conspicuous matches occur (♫229); if you want help listening, focus on the seven time intervals that I provide on the book's website.

To illustrate the marsh wren song-matching game for you with the perfect recording, I spent several exhilarating dawns in a California marsh, learning the behavior and singing locations of two males, finally placing the microphones in just the right place, *all to no avail!* In my stereo recording (♫230) was none of the abundant matching I have often heard (as in ♫227 and ♫229, above). The birds seemed to take turns singing, as if they were listening to each other, but, alas, few if any true matches could be heard. What a wonderful opportunity for future study these western marsh wrens provide for an enterprising, curious naturalist.

More about marsh wrens: "Singing in the Brain" (p. 32).

Explore 28: Song matching by neighboring western male marsh wrens.

Somewhere west of the Great Plains, among western marsh wrens, settle in beside a marsh and focus on the singing of just one male (you can warm up with this recording, almost an hour from one California male: ♫231). What an incredible variety of songs he delivers, each song almost always different from the one before. Careful study would reveal that he eventually repeats himself (as I have illustrated in ♫232), confirming that he has a discrete number of repeatable songs that constitute his large repertoire. Not only does he repeat particular songs, but the song sequences are often the same as well (see ♫233)—maybe that's one way for him to keep track of all his songs.

Now, as with the corvids, expand your listening to hear how two neighbors interact. With a little time and patience, you will begin to hear the matching. Why and when they choose to match is unknown, and sometimes they match hardly at all. Why not? I wish we knew.

Brown thrasher

Here's an example of matched countersinging that I find even more impressive than that of the western marsh wrens. A male brown thrasher sings endlessly until he attracts a mate (p. 28), and in doing so can deliver more than a thousand different songs; and it's likely *far* more than a thousand, because he may well improvise as he sings, or imitate something he just heard from another bird, thrasher or otherwise.

With so many songs, one might think that he sings at random, or out of control, but he is truly the master of his voice, as shown by these sequences in which neighboring males during early spring match each other while sparring on their territorial boundary.

To ease you into hearing the fast-paced replies and counter-replies by two thrashers establishing their territories, listen first in your headphones to these six exchanges (♫234) excerpted from several minutes of sparring by two males. Sometimes the foreground bird follows the distant bird (examples 1 and 5); other times the foreground bird leads (2, 3, 4, and 6). When you are ready, listen for the 49 (!) matches that I hear during the full eight and a half minutes from these two males (♫235).

Then try another example. First, listen to seven excerpted matches from the longer recording (♫236; again, use headphones); notice how the bird on the right track always matches the bird on the left. *Why?*, one wonders. Then try to find those matches and others in the longer program (♫237).

More about the brown thrasher: "Why Sing?" (p. 26), "Small to Large Repertoires" (p. 83), and "The Music in Birdsong—Theme and Variations" (p. 158).

Northern mockingbird

Prowl about in the night and it is extraordinary what you may hear. Just listen to these two mockingbirds have at it on a stage all their own, about 2 A.M., four hours before sunrise, in Los Osos, California (♫238). The bird in the background (right channel, right headphone) begins by imitating a

California scrub-jay. After that it seems to be all mockingbirdese without obvious imitations, but *just hear how the two birds match each other,* playfully (seemingly) hurling the same phrases back and forth at each other! Almost certainly they are two males, but what is their relationship? What do they know about each other, and what are they saying?

More about mockingbirds: "Mimicry" (p. 73), "How Birds Go to Roost and Awake" (p. 108), and "Night Singing" (p. 123).

Oak titmouse

Tufted titmouse, juniper titmouse, oak titmouse—they all engage in matched countersinging, but it's even more striking among the two western species (juniper and oak) given their larger song repertoires. Always, it seems that when you hear one titmouse singing, you can listen in the distance and hear another titmouse singing the same song. In ♫239, listen to three excerpted passages from a longer session, showing how two oak titmice on neighboring

territories match each other. Two mornings spent with these two singing males revealed that they shared at least eight different songs that they had learned from the local dialect.

> ### Explore 29: Song matching by titmice.
>
> (See website for details.)

SONG (AND CALL) DIALECTS

Song dialects occur when a bird learns the unique songs (or calls) at a given location and then stays or returns to sing (or call) there the rest of its life. Any geographical discontinuity, as large as a mountain range or as small as a utility line swath, can help isolate singers of different groups and lead to the development of dialects.

Among species, dialect areas can vary enormously. For some species, two individuals can constitute a mini-dialect, as with chipping sparrows, the indigo and lazuli buntings, or the common yellowthroat (pp. 56–58). For other species, millions of individuals might be involved, such as for the *hey-sweetie* dialect of black-capped chickadees that ranges from Nova Scotia to British Columbia, stretching across the North American continent.

Dialects in human speech develop in the same way, of course, through the relative isolation of people who learn the local speech dialect. Travel the hills and hollows of Appalachia and you'll hear dialects in both birdsong and human speech.

Birds that do not learn songs, such as the flycatchers, do not have dialects, because dialects are, by definition, cultural differences among populations. Songs of these nonlearning species are stable over broad geographic areas, with no local variation, because the song genes are relatively constant everywhere (see pages 45–47).

White-crowned sparrow

The handsome white-crowned sparrow is one of the most celebrated dialect singers in all of North America. The single song of each male typically begins with one or two clear, half-second whistled notes, followed by more complex, rapidly slurred whistles or buzzy vibrato notes. In the examples I provide, you can listen to two dialects from coastal California chaparral (♫240–246), a Montana dialect (♫247–250), and an Alaskan dialect (♫251–254). Or, on your own, walk the chaparral communities along the West Coast (e.g., Point Reyes National Seashore), or hike among meadows of western mountains, and you will soon recognize how dialects of these sparrows change from place to place.

Black-capped chickadee

Throughout most of North America, from the Atlantic Ocean all the way to British Columbia (but not western Oregon and Washington), male black-capped chickadees sing *hey-sweetie* in one of the largest song dialects known (♫255). Smaller dialects occur in isolated populations on the edges of the chickadee's continent-wide range, such as on the island of Martha's Vineyard off the coast of Massachusetts (♫256), and in populations west of the Cascade Mountains in Oregon and Washington (♫257). The males in these examples from the Vineyard and the Oregon dialects are in excited dawn chorus mode (see p. 112), the Vineyard male alternating two different song types, the Oregon male three. To hear the dialect differences back-to-back, you can listen to four songs of each dialect in ♫258.

More about the black-capped chickadee: "Birds Sing and Call" (p. 6) and "The Music in Birdsong—Pitch-Shifting" (p. 155).

Explore 30: Dialects in black-capped chickadee songs.

Everywhere you travel in black-capped chickadee country, listen carefully. Are you in the *hey-sweetie* dialect region? Be sure to get close enough to your singer so that you can hear that slight waver between the two syllables of the *sweetie* note (p. 8); if you are too distant, the reverberation from the *sweet* bouncing off vegetation between you and the singer can obscure that waver. For a special chickadee encounter, experience one of those strikingly different dialects in western Washington or Oregon. Or try those odd dialects on East Coast islands, such as Martha's Vineyard or Nantucket. Listen wherever you travel throughout North America and maybe you will discover yet another local dialect, different from the *hey-sweetie*s, such as someone did in Fort Collins, Colorado.

We are going off topic here, but we take listening experiences where we find them: Listen again to the chickadee and the spotted towhee in ♫257. How often do the two males of these two different species overlap their songs? They mostly alternate, don't they, overlapping far less often than if they were ignoring each other? Each bird clearly seems to be trying to avoid the interference of the other. Or it is possible that one bird is the dominant singer, and the second bird inserts his songs into the silent spaces between the songs of the other. If you are numerically inclined, you could collect some numbers to demonstrate the extent to which overlap of songs is avoided. When attuned to how nearby singing males often avoid interference with each other, you will hear this phenomenon more often. Keep listening!

Dickcissel

Dickcissels sing not just any ol' *dick-dick-ciss-ciss-ciss-ciss* song all day long (p. 29), but rather the one special song of the local dialect. Over distances one can walk in just a few minutes, from one farmer's field to the next, the songs can change: the rhythm, the content, the quality of individual notes, almost everything. Here are examples of three birds from each of three different dialects: from along the Mississippi River, Illinois side (♫259–262); along the Mississippi River, Missouri side (♫263–266); and from Prairie State Park in western Missouri (♫267–270).

More about dickcissels: "Why Sing?" (p. 26).

> **Explore 31: Bicycling among dickcissel song dialects.**
> (See website for details.)

American tree sparrow

Stay alert while traveling and you are likely to hear things no one else has documented. Or even if you are not the first, it is still a joy to discover something by yourself. So it was when I listened along the Denali Highway in Alaska recently and heard a singing American tree sparrow for the first time. A delightful melody of pure tones, I thought, some held steady, some slurred, over two to three seconds. I got to know several birds at mile post 11, only to be thoroughly confused a few miles down the road, wondering what I was now hearing. Turns out that over relatively short distances the dialects of this sparrow change. I provide three examples from along the Denali Highway in Alaska: from mile post 11 (♫271–274), mile post 18 (♫275–278), and mile post 34 (♫279–281).

Fox sparrow

Dialects also occur in the learned songs of fox sparrows. Recognizing those dialects among fox sparrows from the West (the sooty, the slate-colored, and the large-billed forms of this species, p. 145) is a challenge because each male can sing several different songs. In contrast, each male of the red fox sparrow, which breeds from Newfoundland to Alaska, has only one song in his repertoire, and what fun to listen to how that magnificent song changes from place to place. I offer three dialect examples from Alaska, from mile post 11 on the Denali Highway (♫282–284), from mile post 20 (♫285–287), and from Slana, about 65 miles to the southeast (♫288–290).

More about fox sparrows: "Each Species Has Its Own Song" (p. 140).

Vesper sparrow

Vesper sparrow songs everywhere begin distinctively with several clear whistled notes, typically slurred downward, followed by five or so pleasant trills. It is in the relatively simple introductory whistles that one can hear the variation from place to place. Listen to the local dialect of the foreground male and two background males in this dawn recording from the Grand Teton National Park (♫291), where two

introductory whistled notes are slurred downward, followed (usually) by two higher notes, as in *teew teew* ^tee tee^.

Contrast that dialect with one I enjoy in northern Michigan, where the introductory notes are recognizably different, consistently beginning with two or three pure-tone notes, a higher raspy note, and then a low whistle again. Listen to one song from each of three birds (♫292), then to longer sessions from each of the three individuals (♫293, ♫294, ♫295).

I recorded those birds from 2012 to 2015, but I returned in 2018 to listen again, finding that the introductory notes were still the same. Listen to just the introductory notes from nine different males (♫296), or try one complete song from each of those nine males (♫297).

Where vesper sparrows are resident, such as in the Willamette Valley of Oregon, distinctly different dialects in the introductory notes can occur within just a few miles of each other, similar to the noticeable dialect changes over short distances seen there with the resident spotted towhees, Bewick's wrens, and song sparrows (see Explore 26, p. 58).

White-throated sparrow

Although the white-crowned sparrow gets most of the attention when it comes to sparrow dialects, its cousin, the white-throated sparrow, offers its own intrigue. Some common song patterns in this sparrow's songs seem to have disappeared over the years, indicating that songs over widespread areas

can change over time. Although the standard mnemonic for this bird is *ohhhhh sweeeet Can-a-da Can-a-da Can-a-da* (or *poor Sam Peabody Peabody Peabody*), the "standard" *Can-a-da* (or *Peabody*) triplet that once seemingly ended all songs is being replaced with only a doublet, in a transformative wave that began in the West and is pushing east. Odd songs abound in this species, too, suggesting that individuals might be somewhat innovative, or that, unlike white-crowned sparrows, they place no strong premium on learning the same songs as their neighbors, so that small dialects do not seem to exist.

For western doublet songs, I provide examples from Michigan, Minnesota, and Ontario: ♫298–301. Examples of eastern triplet songs are from Maine, Vermont, and Massachusetts: ♫302–305.

Sample "odd" songs are from Ontario (♫306) and Maine (♫307), though each clearly still belongs to its respective tradition, in that the Ontario song has a *Cana* doublet, the Maine song a *Canada* triplet.

More about white-throated sparrows: "Learned Songs of Songbirds, and Babbling" (p. 49).

Explore 32: A road trip through white-throated sparrow song dialects.

Take a road trip, slowly, with the windows open, through white-throated sparrow country, listening carefully to their songs. Start in the East and head west, or reverse the direction. Stop and take a listening sample every 100 miles or so. How far have those western doublets encroached on the eastern *oh sweet Canada* triplet songs? It's a dynamic situation, exciting to track over time, to understand better the puzzling, rapid cultural changes in birdsong that can occur over widespread geographic areas.

Bobolink

The bobolink begins his song with rich, low notes, but soon rises in pitch, higher and higher, accelerating and rollicking upward through what sounds to our ears like an ecstatic jumble of reedy, bubbling notes. These complex songs are tough to grasp with the unaided ear, but if you listen carefully you may hear that most males have two different songs (e.g., ♫308). Listen even more closely, perhaps aided by some recording and analysis, and you realize that his songs are much like the two songs that all of his neighbors are also singing.

Compare here songs of males from two different hayfields in western Massachusetts, one from South Ashfield (♫308–310), the other from Whately (♫311–312). The two hayfields are only six and a half miles apart.

Song dialects in bobolinks can persist from year to year. During 2018, four to six years after I had recorded the above two dialects, I returned to one of the locations (♫313). Can you hear which dialect this male is from?

Last, here's an example of a distant dialect, from Malheur National Wildlife Refuge, Oregon, about 2,300 miles across the continent from western Massachusetts: ♫314.

Anna's hummingbird

Yes, some hummingbirds learn their songs, and the Anna's is a good example! But what a challenge for our ears to appreciate its fast, high-pitched squeaks. Watch the male on his perch, flashing the iridescent rose in his ruffled gorget and crown, and listen to his six-second masterpiece, the components sounding something like *bzzbzzbzz bzzbzzbzz bzzbzzbzz chur-ZWEE dzi!dzi!* and repeated, maybe even for several minutes at a time when he is especially excited. Neighboring males sing alike, but birds only a short distance away can sing a different dialect, as careful study of the songs reveals. Let's listen, hyperattentively, to four birds from three dialects.

DIALECT 1: Male 1, Montaña de Oro State Park, Los Osos, California. Marvel at some of the best songs excerpted from several hours of recording (♫315), but because your ears register so little of what he is up to, listen to one of those songs at full, half, quarter, and eighth speed (♫316). And there's more: At the beginning of his 16 dainty sweet notes all in a row, a pure-toned note stretches above and over the first two notes, revealing that he is using his two voices simultaneously. To hear these simultaneous notes, listen to that series of sweet notes at ⅛, ¹⁄₁₆, and ¹⁄₃₂ speed (♫317), with the higher voice separated into the left headphone, the lower in the right headphone (in songbirds, the low voice is from the left side of the syrinx, the high voice from the right—see p. 38). Better yet, you could import the song into Raven Lite and see these two voices!

DIALECT 2: Male 2 (♫318), Anza-Borrego Desert State Park, Borrego Springs, California.

DIALECT 3: Male 3 (♫319) and male 4 (♫320), Morro Bay State Park, Morro Bay, California (just five miles north of the recordings for dialect 1).

Back-to-back now, testing your ears, listen to summary recordings from all three dialects. The first contains one song at normal speed from each of the four males (♫321). Can you detect how those noisy three- to four-second "scrapes" at the beginnings of the songs differ from place to place? (The last two songs are from the same dialect.) Listen again, and again. Then enjoy these same four songs at half speed (♫322) and at quarter speed (♫323)—I love the details that can increasingly be heard at the slower speeds and lower frequencies. To help you focus on the differences in the songs from place to place, try listening to just the scrapes, back-to-back, from these four males (♫324). To my ears, and to my eyes when I look at the sonagrams, the first three scrapes are strikingly different from each other, representing three different dialects, but the last two are very similar, as would be expected from two neighboring singers in the same dialect.

Red-winged blackbird

The resident males in the Central Valley of California have some stunningly odd, local song dialects, as might be expected of largely nonmigratory birds (see Explore 26, p. 58; also p. 167 for similar song dialects in resident birds of the Willamette Valley, Oregon), but it is the dialects in the red-wing call notes that are especially fascinating. The males within a marsh learn their dozen or so calls from each other, sounding a bit like variations on *check* or *chuck* or *chink,* maybe with a distinctive, loud, down-slurred *tee-yer.* Neighboring males often use the same call simultaneously, as if answering each other, for some reason calling in sync with each other.

I love the distinctive calls from a Colorado dialect, recorded from a male who seemed agitated at my presence, calling continuously for minutes on end (♫325). How different those calls are from the calls excerpted from a singing and calling bout of an Illinois male (♫326). A

Michigan male used about 15 different calls during a 57-minute dawn chorus; listen to a couple of examples of each call type, organized from most complex to least (♫327). If you'd like to know how those calls were used during a 57-minute dawn effort, here's your opportunity: ♫328. There is plenty to explore there besides the male's calls (see pages 18–19, 117)!

While thinking of red-winged blackbird *call* dialects, listen to the *singing* males in this coastal California population (♫329). Can you hear how all of the males (in this stereo recording—again, listen with headphones) seem to be singing the same song, all in tune with the local dialect? You won't hear that kind of unanimity among the more migratory eastern males, where each male improvises unique songs more than do their imitating California cousins.

More about red-winged blackbirds: "Female Song and Duets" (p. 15) and "Energized Dawn Singing" (p. 112).

Explore 33: Call matching by male red-winged blackbirds.

Here's another marsh experience. Settle in and get comfortable, focusing on the calls of one particular male red-winged blackbird. How often does he change from one call to the next? Hear how neighboring males often (how often?) match each other with the same call from their local dialect. Perhaps *when* they change calls might tell you something about *why* they change, or why they have such a variety of calls in the first place. You might have to walk around a bit to keep your bird in a calling mood. Does he change calls in any way that is coordinated with your movements, or your relative threat? Compare your local dialect with the three examples from Illinois, Colorado, and Michigan. Similar? Different? Any unique calls that you have never heard anywhere else?

Brown-headed cowbird

Like the red-winged blackbirds, male cowbirds have dialects in their songs, but because each male can sing several different songs, it is the dialects in their "flight whistles" that are most readily heard, especially in the West. Just before, during, or after a male flies, he offers his brief whistled call, consisting of relatively pure tones that often sound like he is inhaling a squeaky *whsss* and exhaling a whistled *pseeeee.*

Males within a dialect have similar flight whistles, as illustrated by

two males from western Massachusetts (♫330). The flight whistle is often given in flight, of course, but is also given just before the male flies, often after a series of perched songs (e.g., in ♫331, after singing for some time the male calls at 2:05).

The variety of calls among the flight whistle dialects is fascinating, as illustrated by eight sample flight whistles from across the geographic range of the cowbird (♫332): Massachusetts (the two males in ♫330), Virginia, Kentucky (two males), Michigan, Wyoming, and Montana.

More about cowbirds: "Not One But Two Voice Boxes" (p. 38).

Red crossbill

What a curious case the red crossbills offer! As with other species in the Cardueline subfamily (including goldfinches, pine grosbeaks, and more—see p. 176, in "Who's Who?"), the birds learn their call notes from each other. In a mated pair of goldfinches, for example, the male and female have calls more similar to each other than to other goldfinches in the neighborhood, indicating that they learn their calls from each other and converge on a shared call. Among red crossbills north of Mexico, it seems that about eight different regional call dialects have been documented, and professional ornithologists debate whether these call groups might actually represent different species. One of these groups restricted to Idaho has been promoted to full species status: the Cassia crossbill.

DIALECT 1: ♫333. A flock of crossbills at Grand Teton National Park, first landing on the ground in front of us, then flying up to a nearby tree, eventually flying off (0:26) into the distance.

DIALECT 2: ♫334. From Vermont, hear local birders complain first about seeing only siskins ("Aww, all I see is siskins"), then enthuse about the calling crossbills: "Oh, there's one, right here, in the top of the tree, bare tree . . . See it? . . . See it, right on top of this maple, right behind that pine . . . [heavy footsteps] . . ."

MIMICRY

In some songbirds, the singers learn not only the songs of their own species but also the songs or calls of others. The mockingbird is so named for good reason, because it incorporates into its own singing repertoire the songs of so many other species. A number of examples of this mimicry can be found among other song-learning songbirds as well.

Puzzling? Yes! We have no satisfactory explanations as to why birds mimic other species. One possibility is that they do so to acquire a large repertoire of sounds, but the mostly improvising catbirds and thrashers acquire a far larger repertoire than does the mimicking mockingbird, so that explanation doesn't hold. Perhaps the mimics threaten other species by learning their sounds; that could be true of the aggressive mockingbird, but not for most mimics. Maybe jays mimic hawks to scare other birds. Maybe different explanations are required for every instance of mimicry among species, but that's not very satisfying. In the end, it is possible that there is just no good reason at all that some birds mimic. They do it, perhaps because they can, or perhaps because the female "allows" it, for whatever reason.

Northern mockingbird

His name says it all: "the mockingbird." He tells all that goes on in the neighborhood, as nothing seems to escape him while he races from flicker to jay to wren to titmouse to nuthatch to bluebird to whatever else is on his mind. Everything he pilfers is converted to his program, to mockingbirdese, which requires that a sound be repeated a number of times, but someone who knows the sounds in the area can readily pick out his sources.

In seven-plus minutes from a Virginia male, for example (♫335), I can list some of the sources that I hear. For some species the mimicry is superb: the jay calls, the *klee-yer!* call of the flicker, the *wichity* song of the yellowthroat, the *whit-whit* calls of a wood thrush, the *FEE-bee* of an eastern phoebe. You can compare the mimicry by the mockingbird in ♫335 to samples from his model species that occur throughout this book.

0:03 house sparrow
0:30 northern flicker *klee-yer!* call
0:50 purple martin
1:10 blue jay
1:47 common yellowthroat
2:00 common yellowthroat
2:07 great crested flycatcher
2:14 northern flicker *klee-yer!*
2:21 blue jay
2:37 Carolina wren
2:48 northern cardinal
2:56 Carolina wren
3:07 eastern bluebird
3:20 purple martin?
3:29 blue jay

3:47 wood thrush
4:18 northern cardinal
4:27 tufted titmouse
5:01 northern flicker *klee-yer!*
5:04 blue jay
5:12 killdeer?
5:32 eastern phoebe *FEE-bee* song
6:04 great crested flycatcher?
6:16 belted kingfisher?
6:19 wood thrush
6:27 common yellowthroat
6:40 purple martin
6:48 blue jay
6:53 house finch
7:01 tufted titmouse

A second mockingbird from Virginia is an equally fine mimic (♫336). I list a few of the imitations here that are most obvious to me during the first minute:

0:02–0:07 American robin
0:09 American robin (a different call)
0:16 swallow calls
0:17 northern cardinal

0:26–0:29 northern flicker *wik-wik* song
0:38–0:42 blue jay calls
0:42 killdeer
0:48, 0:53 great crested flycatcher

A bird from western Kansas (#337) has altogether different imitations in his repertoire:

Western kingbird (0:00–0:08, 0:41–0:48, 3:43–3:56, 6:49–6:52)
Black-billed magpie (same calls at 0:08–0:12 and 7:03)
Rock wrens—so much material reminds one of rock wren songs (e.g., 0:13–0:31, much of 3:56–4:46, 7:06–7:22, 8:58–9:02)
American robin (0:33–0:40)
Northern cardinals (1:45–1:47, 3:31–3:35, 6:58–7:01, 8:35–8:41)

House sparrow (2:15–2:31, 8:43)
Northern flicker *wik-wik-wik* song? (2:52–2:58)
Great crested flycatcher (4:49)—a single call
Brown thrasher (4:53–5:03)
Red-headed woodpecker (same *kwirr* call at 5:21–5:31, 9:26–9:36)
Northern bobwhite (5:34–5:41)
Killdeer (7:01–7:03)

More about mockingbirds: "Song (and Call) Matching" (p. 58), "How Birds Go to Roost and Awake" (p. 108), and "Night Singing" (p. 123).

Explore 34: Mockingbird mimicry.

Here's another mockingbird you can enjoy (see the website for the mimicry I identified). Day after day he sang from an isolated tree in my neighbor's yard, so one morning I waited in the dark for him to begin. With my parabolic reflector aimed up at him, I recorded three dawn segments, spanning sunrise (♫338, ♫339, ♫340), for a total of 33 minutes. After that, I hoisted a microphone up to his perch and recorded until evening, but provide here only another 27 minutes (♫341).

Here are some highlights from his performance. In ♫338 he creatively alternates calls of the house sparrow and American robin, then calls of the house sparrow and red-bellied woodpecker (6:13–6:19). How he loves blue jays and Carolina wrens! From two singing sessions (♫338, ♫339), I excerpted all the sounds that mimicked the Carolina wren and placed them back to back, in ♫342. Overall, how many species do I hear him imitate? About 20. What a remarkable creature!

What more can I learn from the Carolina wren mimicry? I can estimate how many different songs this mockingbird can sing by extrapolating as follows: (1) The wren mimicry comprises 149 seconds (duration of ♫342) out of the 1,573 seconds in ♫338 and ♫339 combined, or 9.5 percent of all the mockingbird's singing. (2) By studying the sonagrams in Raven Lite, I can identify 9 different Carolina wren imitations, from the 30 to 40 that the male Carolina wren in the neighborhood sings. (3) Dividing 149 seconds by 9 song types, I learn that each of the 9 Carolina wren song types occupies, on average, 16.6 seconds of the mockingbird's effort. (4) Dividing the total singing time of 1,573 seconds by 16.6 seconds, the estimated number of seconds each song type occupies in the mockingbird's effort, I get 95, the estimated repertoire of this male mockingbird. You could try estimating this mockingbird's repertoire based on the stuttered song of the eastern phoebe (occurs three times), the song of the wood thrush (five times), or the song of the northern flicker (six times). If by chance you choose a song that the mockingbird rarely uses, you will overestimate his repertoire size; conversely, choose a favorite song that the mockingbird uses more often than others and you will underestimate the repertoire. The more songs you use the more accurate will be your repertoire estimate.

European starling

I love starlings! This close relative of the mocking-bird is also a master mimic, though so little appreciated. He sings guardedly, it seems, rather softly, and one must be up close to appreciate all that is going on. On occasion, I have made the effort for an intimate encounter, and I am always flabbergasted at what I hear.

Try this song from Michigan (♫343). Overall, the energy and enthusiasm seems to grow until the finale, but the thrill is in the details. His initial down-slurred whistles are typical, but they do not sound quite like the oft-used red-tailed hawk or eastern wood-pewee imitation; from 0:04 to 0:09 are imitations of what sound like an American robin or a hairy or downy woodpecker. Adeptly using his two voiceboxes, he *simultaneously* imitates a flicker's *wik-wik-wik* song and an eastern phoebe's *FEE-bee* song from 0:13 to 0:17. He then offers two sandhill crane bugles, followed immediately by two renditions of a black-capped chickadee *hey-sweetie* song, both the crane and chickadee mimicry occurring *simultaneously* with sounds produced by the other voice box (0:19–0:21). He is truly an extraordinary singer!

Western starlings mimic western species. In ♫344, from Oregon, I hear the following mimicry:

0:01 long-billed curlew (listen to a real curlew in ♫345)
0:04–0:14 California quail: *chi-CAH, chi-CAH, chi-CAH, chi-CAH chi-CAH-go, chi-CAH-go*
0:15 human voice?
0:17 tree swallow
0:19 Pacific tree-frog
0:19 Wilson's snipe winnow

0:22 long-billed curlew
0:25 killdeer
0:26–0:29 house sparrow, sandhill crane; ever so subtly, the two species are imitated simultaneously with the two voice boxes
0:29 American robin song phrase
0:35 Black-crowned night-heron calls?

When listening to these magical singers, I like to remind myself of their origins, as I have often listened to noisy babies in the nest (♫346) and marveled at how those young birds mature into accomplished adults.

White-eyed vireo

CHICK-a-per-weeoo-CHICK! the white-eyed vireo sings, accenting the beginning and ending, gliding more nimbly through the middle. Dissect each of a male's dozen or so songs and one finds that he has incorporated a remarkable variety of call notes from other species—not songs, just the calls.

For the first example (♪347), you might initially hear only the songs of a white-eyed vireo, ten of one kind followed by ten of another. But dissect one of the first ten songs into its components and you hear three blatant mimicries: the call notes of a great crested flycatcher, a Carolina chickadee, and a wood thrush. The mimicry is best heard when the imitated components are isolated and compared directly to the sounds made by an actual flycatcher, chickadee, and thrush (♪348). The last ten songs in this first example clearly begin with the same wood thrush call notes.

In the second example (♪349), listen in the first song for the *klee-yer!* call of the flicker and the *chick-a-dee-dee-dee* call of the Carolina chickadee; the chickadee notes occur elsewhere as well, in a different song (e.g., 3:02). In the fifth song, listen for the single note of a white-breasted nuthatch, and how the vireo squeezes in two notes of a summer tanager's call. It is challenging to hear these imitations in real time, so I have isolated each of them (♪350). It wouldn't surprise me if every one of the notes in these diverse white-eyed vireo songs is imitated from another species, but it is not easy to track down the models for these master listeners and master mimics.

The third example (♪351) also features the calls of a white-breasted nuthatch.

The lesser goldfinch and Lawrence's goldfinch, two western goldfinches, are also known to mimic calls in this way.

> **Explore 35: Tuning your ears to hear mimicry.**
>
> (See website for details.)

Steller's jay

Steller's jays are best known for their loud, harsh calls, such as their jarring *shook shook shook shook shook,* but they have a variety of other typical jay calls as well (e.g., ♪352, ♪353). On occasion, these jays can

fool us with their mimicry, such as with the sliding, asthmatic scream of a red-tailed hawk, *kee-eeee-arrr* (♫354), in this example sounding very much like an abbreviated scream of the red-tail in the field nearby (♫355). And just why do jays mimic hawks like this? Ornithologists have no solid answers, but are prone to speculate, of course. Maybe the jays are trying to scare other birds, or trying to warn their jay companions of a hawk or other danger nearby. Then I wonder, how often do jays call like hawks when we humans are not there to hear them? Or maybe what sounds to us like a hawk is not mimicry at all, but just a normal jay call that by chance sounds like a hawk. (I think it is mimicry, but one should keep in mind all possible explanations.)

Blue jay

Like its close cousin the Steller's jay, the blue jay has its share of typical jay-type calls, but also a variety of others. Listen to the harsh jay calls (to 0:20 in ♫356), then the "squeaky gate calls" (0:22–0:32), and to the "bell" call (0:34–0:42), overlapped with more harsh jay calls at the end.

With the cat having escaped from the house (and me standing nearby), this blue jay imitates a hawk (♫357), or so it certainly seems. Which hawk? The broad-winged hawks call in the woods nearby (♫358), but the jay's imitation is slurred downward, more like that of a red-tailed hawk (♫355). What does it accomplish to call like a hawk at the sight of a cat?

Another blue jay in Florida (♫359) imitated a red-shouldered hawk (♫360), and also gave the "rattle," a call believed to be given only by females. I increasingly wonder if I, an intruding human, cause these hawk imitations.

More about blue jays: "How Birds Go to Roost and Awake" (p. 108).

Other species mentioned throughout the book are also fine mimics: gray catbird, brown thrasher, California thrasher, sage thrasher, Cassin's finch, and yellow-breasted chat.

6. Song Learning Often Creates Complex Songs and Large Repertoires

SONG COMPLEXITY

Among birds, songs of some species are exceptionally simple, others are highly complex, and still others so extraordinary that they can bring tears to the eyes. Yet all living birds can be considered equally "successful," in that each can trace its lineage all the way back to the beginning of time. Why simple songs are sufficient for some species but not others is yet another mystery of birdsong.

Both flycatchers and songbirds are featured in this section. The flycatchers have innate songs, and therefore have relatively simple songs and small repertoires. Songbirds learn their songs, and while some of them puzzlingly restrict themselves to simple songs and small repertoires, it is only songbirds that have the highly complex songs and the larger song repertoires.

Songs that are a simple burst of sound

Least flycatcher

Che-BEK! says all that the least flycatcher knows in just an eighth of a second (♫361, ♫362). Only a single song, but how emphatically he delivers it, his entire body from head to tail thrown into each split-second effort. And how insistent, delivered during the dawn frenzy at a clip of more than one a

second, as if trying to outpace all of his neighbors who are doing the same.

More about the least flycatcher: "Song Changes over Evolutionary Time, from Species to Species" (p. 147).

House sparrow

Cheep, chirrup, churrp, and the like, endlessly, each only a quarter second, a singing house sparrow in peak form delivers them at a clip of almost two per second. To even the trained ear the songs can all sound alike (♫363), and sometimes they actually are, as revealed when we slow those songs to half

speed (♫364). Other times a male delivers a greater variety of songs, detectable to the unaided ear (♫365), but especially apparent when slowed to half speed (♫366). Surprisingly, a hungry nestling house sparrow seems to "call" with all of the variety that an adult male "sings" (♫367; slowed to half speed, ♫368). How odd, as I know of no other songbird like this; in typical songbirds, the nestling uses calls nothing like the song of the adult, and the youngster has an extensive period of song practice (see p. 50) before acquiring anything like the adult song.

Songs that consist of one repeated phrase

Cactus wren

The state bird of Arizona, the voice of the desert Southwest . . . the cactus wren's song creeps gradually into consciousness, *char-char-char-char-char-char-char-char-char-char-char.* The raspy, mechanical repetition of a single phrase builds throughout, though by wren standards he always sounds a bit subdued. He repeats himself several to many times, then switches to a different song, *rar-rar-rar-rar-rar-rar-rar-rar-rar-rar-rar-rar-rar-rar,* and eventually

others, though no one knows for sure how many variations he holds in his head.

In ♫369, listen to the first 14 songs, all the same and no doubt by the adult male on the territory. He continues with the same song throughout the eight minutes, but a second bird chimes in twice (1:59, 2:41). Because the territorial adult male does not attack this occasional singer, he could well be a young male still on the territory and soon to disperse. That song at 2:41 also has a slight wavering quality to it, again suggesting the singer is not a full adult.

Dark-eyed junco

The song of the dark-eyed junco is often con-fused with that of a chipping sparrow, but there is usually a lingering ring to the song of a junco, a tinkling, musical, slightly quavering quality that only the junco knows. He typically repeats himself, often over 100 times, before switching to another of the several songs he knows, but during the dawn chorus (as in ♪370) he excitedly alternates his songs. How many different songs do you hear? In what sequence does he sing them? Enjoy the great horned owl who has not yet gone to roost for the day.

More about juncos: early spring song, in Explore 25 (p. 55).

This same relatively simple song construction occurs in several other species mentioned in this book: loggerhead shrike (p. 160), oak tit-mouse (p. 63), tufted titmouse (p. 87), white-breasted nuthatch (p. 9), rock wren (p. 90), Carolina wren (p. 17), chipping sparrow (e.g., p. 56), clay-colored sparrow (p. 122), ovenbird (p. 138), and common yellowthroat (e.g., 58).

More complex songs contain several different phrases

Among the many species that occur throughout this book, I offer a focus on two.

Painted bunting

The painted bunting is *nonpareil* in French, meaning "without an equal," an apt descrip-tion for the stunning plumage of the males who are two years or older. His songs are a pleasant sequence of about ten different phrases, jumping abruptly from high to low frequencies over two seconds, but overall his songs feel somewhat thin and weak, not heard at any great distance (♫371). One effort to capture his song in a mnemonic is *tida dayda tidaday teetayta totah*.

House wren

"*O-du-na'-mis-sug-ud-da-we'-shi*," which translates to "Big Noise for Its Size," is how the Chippewa named the saucy, tireless house wren. He stutters a few chattered notes at first, then suddenly rises to a higher pitch, louder now, bubbling and gurgling his ecstatic song down the scale, a gush of energy that is soon repeated—or given continuously for a minute or more, should a prospective mate be near (♫372).

Exceptionally complex songs from exceptional singers

Then there are the truly exceptional singers, those with hundreds or even thousands of sound bits and pieces all strung together, sometimes taking a minute or more to deliver a single song. "Without breathing, for an entire minute or more?" you might ask. No, in the rapid-fire world of songbirds, they can take mini-breaths at the slightest pause between the notes in their songs. In song, these birds are *nonpareil*.

Townsend's solitaire

Those who hear the male solitaire's song gush about it in superlatives, calling it the most beautiful, the richest and fullest and clearest and sweetest and sparklingest, the finest and most glorious mountain music imaginable. A truly versatile singer, he strings together untold numbers of brief, warbled notes into a song lasting half a minute or more (♫373, with coyotes); marvel at the intricacies when slowed to half speed (♫374).

In the roaring wind of the Cascade Mountains, this second solitaire countersings with another in the ponderosa pine treetops (♫375). Hear that "*three-beers*" at 0:21? Sure sounds like an imitation of an olive-sided flycatcher (♫376), doesn't it? Hear the *three* without the *beers* at 1:18? And never again do you hear those two sounds in all the singing by this bird. Those little hints tell how he splices the bits of his song together, and also hint at how many different bits are in his head, with so little repetition occurring during these minutes of singing. The call note at the end is typical solitaire (from 3:21 to end).

Sage thrasher

Heard best a few hours before sunrise, when all others in sage country are silent, he bursts into song, releasing just a few seconds of what sounds like pent-up boundless energy, rollicking and frolicking, but he abruptly stops, offering only an abbreviated song. With the advancing light his effort grows, and by daybreak his songs last a minute or two or more, a torrent of hundreds of different, half-second sound bites that leaves a human listener breathless, as well as ecstatic, inspired, and just plain happy (♫377). You can appreciate his songs all the more at half speed (♫378). And embedded in all that ecstasy is mimicry of other species, such as western meadowlarks, Brewer's sparrows, soras, gulls, nighthawks, and the like.

More about the sage thrasher: "Singing in the Brain" (p. 32) and "Night Singing" (p. 123).

> **Explore 36: Scaling song complexity.**
>
> (See website for details.)

SMALL TO LARGE REPERTOIRES

Some birds have stored in their brains just one song, others hundreds or even thousands. Why some birds need so many songs while others are content with so few is yet another mystery of birdsong.

Small repertoires of nonlearning flycatchers

Among flycatchers, which do not learn their songs, repertoires are relatively small, as if it is difficult to encode very many songs into the genes.

Olive-sided flycatcher

Hip! THREE CHEEEERS! (or *Quick! Three Beers!*) sings the olive-sided flycatcher. The *Hip* is sharp and brief, the *THREE CHEEEERS* slurred whistles each a half second long, the *CHEEEERS* descending and drawn out. He has just one song that he uses during the dawn chorus and throughout the day, so different from his close relatives the wood-pewees (in the same genus *Contopus*), who have one special song that they use during the dawn chorus (p. 148). Here are three examples, illustrating essentially the same song and same song genes from Michigan to Colorado to Alaska, with a comparison of energized dawn singing (♫379) to more leisurely singing during the daytime (♫380, ♫381).

Black phoebe

Like the eastern phoebe (e.g., p. 46), the black phoebe uses two songs, often described as *tee-hee* and *tee-hoo*. These two songs are often alternated (♫382), whether during excited dawn singing (0–2:20 in that recording) or during a slower pace after sunrise (2:30–4:30).

Say's phoebe

What fun listening to the two or three distinctive songs of the Say's phoebe. By day (♫383), he sings *pit-TSEEeeuur,* distinctively down-slurred and instantly recognizable, and after a few songs he seems to ask the question, *prrrREEP,* rising and raspy. How curious that, in energized singing during the dawn chorus (♫384), this phoebe adds another song, unlike other phoebes but much like some other flycatchers, such as the eastern and western wood-pewees (p. 148). This third song, *pit-urr-EE-EEP,* also rises but is more pure toned, not raspy. Listen for yourself to ♫384 and hear how he

offers all three song forms in the first five songs: *pit-TSEEeeuur . . . pit-TSEEeeuur . . . prrrREEP . . . pit-TSEEeeuur . . . pit-urr-EE-EEP.* Test your ears and try to chart his progress during the remaining hundred-plus songs.

In a third example (♫385), the singer also uses all three songs, but with less vigor, and as a result he seems in an entirely different mood from the energized dawn singer in ♫384. Perhaps not surprisingly, the singing moods of these three birds reflect when they were recorded: 35 minutes before sunrise (♫384), ten minutes after sunrise (♫385), and late morning (♫383). (See pages 113–122 for examples of how singing gradually or abruptly becomes less energized after the dawn chorus).

Other flycatchers mentioned elsewhere in this book also have limited song repertoires: the alder flycatcher (p. 45) and least flycatcher (p. 79) have just one song apiece, the western wood-pewee (p. 148) and eastern phoebe (e.g., p. 46) two songs, the eastern wood-pewee (p. 148) and willow flycatcher (p. 45) three songs apiece. Many of the close relatives of the alder and willow flycatchers in the genus *Empidonax* have just two to three songs apiece (p. 151).

> ## Explore 37: How phoebes use their two to three songs.
>
> Merely identifying the two songs of the black phoebe, the two of the eastern phoebe, and the three of the Say's phoebe is not enough, of course. Once you have identified the songs, it's inevitable that you will ask yourself how they are used in a performance. After you have explored the next section, "How a Repertoire Is Delivered," you will listen more carefully whenever you encounter a phoebe in nature. For the eastern phoebe, you will probably hear that the faster he sings the more likely he is to alternate his two song forms. You can check that: Collect some numbers from time to time on singing rates and how he uses his two song forms, to see what *your* eastern phoebes do. Only the Say's phoebe has a special song used at dawn. How does he work that into the sequences of the other two songs? Does the black phoebe ever sing strings of one song, followed by just a single example of the second, like the eastern phoebe? Phoebes, even with their limited repertoire of songs, can captivate the listening ear.

Just one song for some songbirds

Unexpectedly, given the great potential for song complexity and large repertoires among the song-learning songbirds, males of many species have just one song, and sometimes a very simple song, that is repeated over and over, with relatively minor differences among renditions.

Ruby-crowned kinglet

Each ruby-crowned kinglet has just one song, and he doesn't even always complete it (♫386). He begins high and thin in a squeaky voice, squeezing out two or so seconds of *tseee* notes that gradually shorten; he then plummets to lower *tew tew te te* notes, which also shorten and drop in frequency, before ending with a bold surprise, a galloping, emphatic *tee-LETT*

tee-LETT series (or a three syllable *tee-da-LETT*). Put it all together and it's something like *tseee tseee tseee tsee tsee tse tse tew tew tew tew tew tew te te te tee-LETT tee-LETT tee-LETT tee-LETT tee-LETT.* Feel the different mood in the second bird (♫387), who sings primarily partial songs and adds several upsweeping notes between songs.

> **Explore 38: Songs of ruby-crowned kinglets.**
> (See website for details.)

Savannah sparrow

The male savannah sparrow sings a lazy, high-pitched, buzzy, insectlike song, beginning with soft, stuttered notes for a second or more, followed by two fine, drawn-out, high buzzy notes, the first higher than the second: *tsit-tsit-tsit-tssit-tssit-tssit-tseeeeeeeeeeee-tssayyyyyyyyyy,* or *take take take take-it eeeeeeeaaasssssssyyyyyyy.* Resident, nonmigratory birds can have local dialects, as in coastal California, and the general form of the song can vary geographically, as in these examples from Michigan, Colorado, and Alaska (♫388, ♫389, ♫390).

Other songbirds that typically use only one song throughout the day include the following:

Wrentit (p. 18)	Chipping sparrow (p. 56)
American tree sparrow (p. 66)	Henslow's sparrow (p. 41)

White-throated sparrow (p. 67)

White-crowned sparrow (p. 64)

Ovenbird (p. 138)

Tennessee warbler (p. 53)

Connecticut warbler (p. 128)

Common yellowthroat (p. 58)

Magnolia warbler (p. 54)

Lazuli bunting (p. 57)

Indigo bunting (p. 57)

Dickcissel (p. 65)

Two to four songs is, for other songbirds, just a little better than one

Tufted titmouse

In addition to making a variety of fussy, scolding notes, the tufted titmouse sings, and during early spring he sings nearly all day long. Most titmice songs consist of a repeated, distinctly two-syllabled phrase, as in *peter-peter-peter;* others are one syllable, as in *here-here-here,* or a drawn-out *heeere-heeere-heeere.* Each male has three or four favorite songs that he sings throughout the day, though you must be patient to hear them, given how many times he may repeat one before switching to another.

From Indiana, a singing male offers two of his song types (♪391). Listen also for neighboring tufted titmice to match each other, much as oak titmice do (p. 63). In ♪392, the foreground male sings the same song throughout; the male in the background offers a dissenting voice at first (0:03), but on the next song matches the song of the foreground bird. (Another example of matching tufted titmice occurs in ♪447.) Listen also for how songs change from place to place, in the form of dialects (p. 63).

Beware: For me, the tufted titmouse is one of two species (the other being the eastern towhee) who sometimes deliver a song so atypical that I have no clue as to who sang it.

Carolina chickadee

Songs of the Carolina chickadee consist of brief, pure-toned notes, high-pitched and therefore sharper and more piercing than those of its black-capped cousin (p. 155). He typically uses four notes, often in a high-low-high-low pattern, such as *fee-bee-fee-bay,* or in other patterns, such as high-low-low-high, *fee-bee-bee-fee.* During the day, the

male sings one of his two or three songs many times before switching to another; at dawn, he cycles among them more quickly.

In this typical sequence (♫393), the chickadee repeats a song several times before switching to another, occasionally alternating two different songs. In ♫394, try tracking how a male uses his different songs during a quarter hour. In ♫395, recorded during the dawn chorus, the male switches among his three songs (A, B, and C) more frequently, singing in the following pattern: A A A A B B B B B B A B A B A B A A C C C C C B (with occasional "gargle calls" interspersed—see p. 8).

Other songbird species in which males sing just a few songs include the following:

Philadelphia vireo (p. 98)
Tree swallow (p. 132)
White-breasted nuthatch (p. 9)
Winter wren (p. 166)
Veery (p. 10)
Gray-cheeked thrush (e.g., p. 128)
Swainson's thrush (p. 161)
House finch (p. 31)
Purple finch (p. 161)
Eastern towhee (p. 94)

Cassin's sparrow (p. 96)
Field sparrow (p. 121)
Fox sparrow (e.g., 145)
Dark-eyed junco (p. 81)
Yellow-headed blackbird (p. 30)
Bobolink (p. 68)
Red-winged blackbird (e.g., 117)
American redstart (p. 139)
Chestnut-sided warbler (p. 139)
Black-throated blue warbler (p. 28)

Songbirds with ten or so different songs

Hermit thrush

A gifted songster, the inspiration of countless poets, the hermit begins his song with the purest of tones, stretching to nearly half a second, then whirls and twirls through a fluty flourish for a full second to finish his masterpiece: *ooooooooooooooooh, holy holy, ah, purity purity, eeh, sweetly sweetly* is how you might hear it. But listen to the next song, and it will

be different, and the next, with the male often offering all ten or so of his different songs before beginning all over again. Appreciate the variety of songs in the four examples here, from Massachusetts to Virginia to Arizona, and do your best to identify a particular song (maybe an especially high one) that you will hear him repeat about every ten songs (♫396, ♫397, ♫398, ♫399).

More about the hermit thrush: "The Music in Birdsong—Contrast" (p. 156).

Explore 39: Seeing the magnificence of hermit thrush songs.

Even more satisfying than listening to a hermit thrush is to both listen and see his songs simultaneously. Import one of the hermit thrush recordings into Raven Lite. Then pick a unique, recognizable song and *see* how it resurfaces about ten songs later. If you print 20 to 30 sonagrams and sort them, you will come up with about ten distinctly different piles, indicating that each of the male's songs is well practiced and committed to memory. While in Raven Lite, play the songs at slower speeds, and backward, too.

Northern cardinal

There is so much to learn from northern cardinals; here we focus on the variety of their songs. Each individual (recall that both males and females sing, p. 19) has about a dozen different songs, and each song type is typically repeated many times before switching to another. Listen to the male in ♫400 switch to a new song several times, at 0:45, 2:14, 3:30, 5:37, 6:06, 6:44, and 7:36. Then try your increasingly refined ear on the other three examples

as well, from Virginia (♫401), Kentucky (♫402), and Florida, beside the Gulf of Mexico (♫403).

The closely related pyrrhuloxia sings in much the same way, repeating one song several times before switching to another (♫404).

More about cardinals: "Female Song and Duets" (p. 15), "Not One But Two Voice Boxes" (p. 38), "Song (and Call) Matching" (p. 58), and "How Birds Go to Roost and Awake" (p. 108).

Some other species in which males are known to have about ten different songs include the following:

White-eyed vireo (p. 77)
Bell's vireo (p. 164
Oak titmouse (p. 63)
Pacific wren (p. 42)
Bewick's wren (p. 167)
Spotted towhee (p. 94)

Eastern towhee (resident Florida
  birds; p. 94)
Song sparrow (p. 51)
Red-winged blackbird (p. 117)
Western meadowlark (p. 151)
Brown-headed cowbird (p. 40)
Yellow warbler (p. 115)

Explore 40: Getting to know song repertoires of individual birds.

For any species with a smallish repertoire, listen to an individual over an extended period and try to verify for yourself that the singer has only one or a few different songs. Here's a satisfying venture: One by one, get to know by ear all ten or so songs of a bird, like a song sparrow. Once you know all of his songs, you can follow along, hearing which song is on his mind through whatever and whenever circumstances arise. All kinds of questions then follow about how and when and where he uses his different songs, and you can begin to answer your questions.

Songbirds with hundreds of songs

Rock wren

Listen easily to song after song of the rock wren. Each is a buzzy or ringing phrase repeated several times, maybe a second and a half in all, and every five seconds comes another, with successive songs remarkably different from each other. *Chair chair chair chair . . . karee karee karee . . .* But now listen more closely, as he may alternate two different songs for a half minute, then introduce a couple of others, gradually revealing in this fashion all that he knows.

During the first minute in ♫405, for example, the male sings a dozen songs of five different types, in this sequence: A B A C B C B D C B D E. For fun, I checked where song D occurred again: Beginning at 16:18, about 200 songs later, he offers four renditions of D alternated with another song. During an intense dawn chorus, males sometimes call as much as they sing (e.g., from 0:06 to 0:12 in ♫406).

Sedge wren

A sedge wren improvises his large repertoire (p. 48), but the songs are so similar to each other that it is often difficult for us to detect exactly what he is up to. Listen to a sedge wren as you would hear him in his sedge meadow (♫407), delivering 36 songs over about two minutes. He switches to a new song at 0:25, 0:46, 1:21, and 1:52. Listen to one example of each of the five different songs he sings (♫408), then to those same five songs slowed to half speed

(♫409). Last, try listening to just the ending trills at quarter speed (♫410); now we begin to hear the variety in the songs that the birds themselves hear in their hundreds of distinctive songs.

More about sedge wrens: "Improvised Songs" (p. 47), "Each Species Has Its Own Song" (p. 140), and "The Music in Birdsong—Improvisation" (p. 159).

A few other species with a hundred or so discrete, identifiable songs include the following:

Marsh wren (western; p. 60)   Northern mockingbird (p. 73)
Gray catbird (p. 49)

And songbirds with more songs, thousands even

There are a few truly exceptional singers who do not have a discrete repertoire of different songs but instead have unlimited potential to produce new and different songs as they sing. The one individual sage thrasher I studied, for example, was estimated to have about 700 different, half-second song units stored in his head, and by rearranging them in his long, complex songs he could produce an endless variety (p. 83). The Townsend's solitaire probably generates infinite variety in the same way (p. 82). The brown thrasher may actually improvise songs as he sings, or copy instantaneously the sounds that he hears from another thrasher or another species, thereby generating endless variety in his performance (p. 92).

Note that here is a second way in which we might think of songbirds improvising songs. The first was in how a bird acquired a sound in the first place, making it up, or improvising it, as opposed to learning it by imitating another individual (p. 47). Once a bird such as a sage thrasher has acquired his 700 or so sounds, he then might recombine them to produce new sequences—that is, he might improvise new songs based on the sound units that he already knows.

Brown thrasher

Oh my, here is the ultimate listening experience, using the mimicry by a brown thrasher to estimate by ear how many different songs he sings. Let me lay it out for you. In these seven recordings of a brown thrasher are three hours and 15 minutes of concentrated listening (♫411–417), with the recordings arranged by decreasing quality (not by date), and each recording having had some longish silent intervals removed. I gathered these recordings because I yearned for this experience myself, and it took me about five years to find a bird that would cooperate long enough to give me the recordings I had hoped for (I'd have had many more hours recorded from him if a female hadn't arrived, whereupon he quit singing—see p. 28).

As I listened to these recordings, I searched for "handles" on his versatility, particular song phrases that I was confident I would recognize by ear if they occurred again. I settled on ten such songs, all believed to be imitations of common birds. I extracted those ten from the thrasher's program and play them back-to-back in the following sequence (♫418):

1. *hey-sweetie* of black-capped chickadee, abbreviated to a quick *hey-sweet* (nine, the number of times this chickadee song occurred in the thrasher's three and a quarter hours; so, on average, the chickadee imitation occurred once every 22 minutes, which might be called the "return time")

2. a hurried *pee-a-wee* of an eastern wood-pewee (two times, occurring on average once every 98 minutes)

3. tremolo of common loon (eleven times, 18 minutes)

4. song of eastern meadowlark (one time, 195 minutes)

5. call of eastern meadowlark (two times, 98 minutes)

6. song of chipping sparrow (eleven times, 18 minutes)

7. song of northern cardinal (seven times, 28 minutes)

8. song of Nashville warbler (seven times, 28 minutes)

9. *flicka* call of northern flicker (two times, 98 minutes)

10. *che-BEK!* song of least flycatcher (eight times, 24 minutes)

To illustrate this mimicry, I found examples of the ten model songs from actual chickadees, wood-pewees, and so on, and play each on the left track, with the thrasher's imitation following immediately on the right track (♪419). Many of the imitations are spot-on, unmistakable. Someone might suggest that, because a thrasher has so many different songs, by chance alone he is likely to sound as if he were imitating. Yes, I would counter, but then you would expect mimicked sounds of western species in his repertoire, too, but in reality he "imitates" only those species that he is likely to have encountered during his travels in the East. Some of the model singers had not yet arrived on the breeding ground when this thrasher was singing, so from previous encounters, perhaps the previous year, he must have remembered the songs of the wood-pewee, Nashville warbler, and least flycatcher. Alternatively, I suppose he could have learned those songs from another thrasher that spring, but then that thrasher would have had to remember those songs from previous encounters.

Here's the experience I was after: Can I estimate by my unaided ear how many different songs this thrasher sings? *Yes,* I can! I concentrate, counting how many "doublets" (i.e., "songs"—see p. 158 for how a brown thrasher sings in doublets) a male sings in a given recording, simultaneously keeping track of how many times I hear each of the imitations. My tallies for each of the ten imitations are in parentheses above, after each of the species' names. These particular song phrases, the "handles" by which I could pry into the mind of this bird, occurred a total of 60 times (9 + 2 + 11 . . . + 8 = 60), on average 6 times apiece (60/10 = 6). Altogether I counted 7,697 songs (the doublets); 7,697 divided by 6 = 1,289. That's about 1,300 *different* songs that this male brown thrasher sang during the three and a quarter hours I listened to him.

One take-home fact from this exercise is that a brown thrasher does have discrete, learned songs that are stored in his brain and that he can retrieve when he chooses. What we cannot know is if this thrasher improvised novel songs during his performance; to the extent that he did improvise, the 1,300 is an underestimate.

More about the brown thrasher: "Why Sing?" (p. 26), "Song (and Call) Matching" (p. 58), and "The Music in Birdsong—Theme and Variations" (p. 158).

> **Explore 41: Patience in listening to songs of a brown thrasher.**
>
> (See website for details.)

HOW A REPERTOIRE IS DELIVERED

When a bird has at least two different songs stored in his brain, he begins to have choices as to how to deliver them.

Birds who typically sing with "eventual variety"

Eastern towhee, spotted towhee

The standard mnemonic for the eastern towhee, a large, truly handsome sparrow, is *drink-your teeeeeeee,* two or three distinctive introductory notes followed by a series of repeated phrases. Sometimes the repeated phrases are hurried into a buzzy trill, other times repeated so slowly you can easily count them. Listen to this bird in ♫420 repeat one of his songs many times over. If you are patient and a male continues to sing, eventually you will hear him switch to another of his three or four different songs—in other words, he sings with "eventual variety."

Then listen to a male dazzle at dawn (♫421; to the left in the background is a scarlet tanager also in dawn song). You will hear him reveal his entire repertoire far more quickly, often alternating two or even three different songs or bits of songs, perhaps interspersed with his typical call note, *chewink.* (Southeastern, nonmigratory eastern towhees in Florida have repertoires of seven to nine songs, often with simpler introductions, and a call note that sounds more like *sweee*.)

The closely related spotted towhee of the West behaves in much the same way. During relaxed, non-intense singing, he repeats one of his five to nine songs many times before switching to another. In ♫422, the male offers over 150 renditions of the same song, never switching to another during the entire 18 minutes. More variety can be heard in ♫423; listen for the strikingly different song type introduced after 5 minutes. How many other song types do you hear during the 26 minutes? As with

the eastern towhee, during the dawn chorus different song types are often alternated, and he sings fewer renditions of each before introducing another; in ♫424, listen to the male awake with *zhree* notes (hear the Swainson's thrush awaking in the background), then offer three different song types over four minutes, often alternating two song types (i.e., A B A B A B).

Black-throated sparrow

Out of the desert Southwest comes the delightful, delicate song of the black-throated sparrow. Stand with him (♫425) as he awakes and begins to sing in muted tones, as if warming up to the day. Over and over he repeats his song (say, A), until he fades into a pause (from 1:22 to 2:15), then offers his song A some more, pauses again, then resumes at 5:47. Now, however, as if the warm-up is finished, he alternates two different songs, A and B. Next he introduces C (at 10:41), and after another pause brings on D (at

12:49), then E (13:00); you can explore by yourself beyond that point. What fun to follow along with him, listening to all 22 minutes of his dawn singing, to hear how he repeats, then alternates, introducing new songs along the way.

Questions abound. How many different songs does he use? Why repeat, and why alternate? Is there a pattern as to when he does what? Do black-throated sparrows always awake in repeat mode, then after a warm-up alternate songs more? Eastern towhees often do that at dawn.

Bachman's sparrow

From the open piney woods of the Deep South, the songs of the Bachman's sparrow are unforgettable; he is arguably the finest songster among North American sparrows, ranking up there with some of the thrushes for the best of the best in North America. His typical songs are two-parted, consisting of a half-second pure tone note (sometimes buzzy) for an introduction, much like the hermit thrush, followed by a second-long trill of repeated phrases. With about 30 to 40 variations on this theme, during relatively relaxed daytime singing he often chooses to deliver two or three renditions of each before moving on to the next. At other times, perhaps especially during the dawn chorus (e.g., in ♫426), he seems in more of a hurry, racing through his repertoire and offering a second rendition only about half the time. Listen to the first three minutes of ♫426 and you'll hear the following program: A B C D E E F G G H H I J I K L M M N N.

Cassin's sparrow

From the arid shrub grasslands of the southern high plains sings the Cassin's sparrow, a close relative of the Bachman's sparrow. Lacking decent song perches, he often flutters about 15 feet into the air, then sings as he glides back to earth. It is a stunning sight to watch, especially with him silhouetted against the brightening sky before sunrise. This sparrow programs his efforts much like a Bachman's sparrow does, repeating a song just a few times before introducing the next. In ♫427, during the first three minutes, this sparrow sings A A A B C C C B B B C C C A C B B B B B C (songs A and C are wickedly similar to each other). Curiously, each male has just a few songs in his repertoire—three is typical. Listen more closely (e.g., at 1:02) and you hear how the background Cassin's sparrow sings the same song as the foreground focal bird; the matching is again evident later, from 3:08 to 3:17.

Other examples of birds who typically sing with "eventual variety" include the following:

Loggerhead shrike (p. 160)

White-eyed vireo (p. 77)

Carolina chickadee (p. 87)

Black-capped chickadee (p. 155)

Oak titmouse (p. 63

Tufted titmouse (p. 87)

White-breasted nuthatch (p. 9)

Canyon wren (p. 154)

Pacific wren (p. 42)

Winter wren (p. 166)

Sedge wren (p. 91)

Carolina wren (p. 17)

Bewick's wren (p. 167)

Cactus wren (p. 80)

Song sparrow (p. 51)

Dark-eyed junco (p. 81)

Eastern meadowlark (p. 150)

Western meadowlark (p. 151)

Red-winged blackbird (p. 117)

Northern cardinal (p. 89)

Pyrrhuloxia (p. 59)

Explore 42: Once you are attuned, *every singing bird* becomes interesting.

Once you are attuned to the different songs that a male can sing, *every singing bird* becomes interesting. For those birds who routinely sing with "eventual variety," you will be curious to know how many times a male repeats a given song type before switching to another, and how the patterns might change throughout the day, from the first songs of the morning during the dawn chorus until well after the sun is up. Just how does the male vary his presentation to express himself? As you listen, you may come to agree that the more intense the singing, the more likely it is that a male will switch to "immediate variety," with successive songs all different, as with the eastern towhee, spotted towhee, and black-throated sparrow examples above (see also "Energized Dawn Singing," p. 112). Collect some numbers and quantify what you hear. How many songs does he sing each minute, for example, and how many times does he switch from one song type to another? Try plotting your numbers on a graph, with "singing rate" and "percentage of different successive songs" as your axes, and you will begin to see how he programs his performance (see pp. 118–119 for sample graphs for the red-winged blackbird).

Red-eyed vireo

Vireo, sings the red-eyed vireo, with so many nuanced declarations (p. 98), but listen once again for an odd song that will be recognizable when he repeats it, then count how many other songs he delivers until you hear that unique song again. About 15? 18? 20? If you can iden-tify a second unique song, try counting with

that song. Because a red-eyed vireo tends to sing most of his songs before he begins another round through his repertoire, the numbers allow you to estimate roughly how many different songs he knows. (It certainly helps to see the sonagrams from a recording, but you can do this by ear alone in the field too.)

As I listen to the vireo in ♪428, for example, I pick out three high notes at 0:06 and begin counting . . . 29 to the next one at 0:41. If I choose the triple down-slurs at 0:11, I count 8 songs to the next one at 0:22, then 18 to the next at 0:45, though it is only a double note here. That distinctive note at 0:30: I count to 59 for the next one (at 1:43), but then it's only 24 to the next occurrence (2:13). Clearly this male is not too regular in his programming, and he may have favorite songs. But do enough counting, take the middle value (i.e., median) of your efforts (29, 8, 18, 59, 24; middle value 24), and you will have a rough estimate of how many different songs he sang while you listened.

I offer a second red-eyed vireo (♪429), this male a greater challenge as he sings in super-high speed; he is also a greater challenge because he has few really distinctive songs to grab onto and listen for a repeat later.

More about red-eyed vireos: "Improvised Songs" (p. 47) and "Learned Songs of Songbirds, and Babbling" (p. 49).

Philadelphia vireo

The little-known Philadelphia vireo of hardwood forests across Canada is a fascinating puzzle. He looks most like a warbling vireo but sings like a red-eyed vireo, so much so that distinguishing the Phil-adelphia and red-eye by their songs is difficult both for humans and for red-eyed vireos. It is possible that the smaller Philadelphia vireo uses his songs to defend a territory not only against other Philadelphia vireos, but also, in some circumstances, against red-eyed vireos.

In the following examples, first appreciate how similar the songs of the Philadelphia vireo are to those of the red-eyed vireo. Then, as with the red-eyed vireo (see p. 98), pick a unique song and start counting until you hear that song recur. In ♫430, try the first song: I count to 5, then 4 to the next recurrence, 6, 4, 6, with a middle value of only 5 to the next rendition of that same song. In a fit of fascination, I study the sonagrams, documenting his one minute of songs in the following sequence: A B C D E A B D C A B D B E C A D B F A D B C D F A C B E B D (there seem to be three slightly different variants of the B song). For the most part, five different songs (A B C D E) are used, with two examples of a sixth (F) thrown in.

Were I to stop here, with that one recording, I'd conclude that the Philadelphia vireo sings much like the red-eyed vireo, but with a smaller repertoire. But I continued. In another recording made in the same general area (♫431), try counting with the first song: 2, 3, 3, 3, 3. Study all of his sonagrams and you see that they are programmed like this: G H G̲ I̲ H̲ G̲ I̲ H̲ G̲ I̲ H̲ G̲ I̲ H̲ G. He favors the G I H song triplet as a repeating series (underlined).

Now try the third example, recorded nearby (♫432). Why, it's the identical program as for the second recording (♫431), so different from that of the first (♫430)! Again, I program his sequence of songs: G̲ H̲ I̲ G̲ H̲ I̲ G̲ H̲ I̲ G̲ H̲ I̲ G̲ H̲ I̲ G̲ H̲ I̲ G̲ H̲ I̲ G I H G I. He now favors the sequence G H I (underlined), not G I H, but the last four songs reveal that he is flexible as to the sequence he chooses!

What is certain in all this is that I now know how to distinguish the similar songs of the Philadelphia and red-eyed vireos. I simply memorize one particular unique song and then count to its next occurrence. Numbers less than 10 are a Philadelphia vireo, greater than 10 a red-eyed vireo.

How many singing males are involved in my three recordings? One possibility is that two individuals are involved, each with very different song repertoires; ♫430 is from one bird, ♫431 and ♫432 from the second bird. If, however, a male Philadelphia vireo sings one group of songs for a while and then switches to another "package" (see next section), as do some other vireos, it is possible that only one bird is involved. Because vireos aren't renowned for their song learning, I reject a third possibility, that the identical songs in ♫431 and ♫432 are from two different males.

> **Explore 43: Song packages of the Philadelphia vireo.**
>
> (See website for details.)

Other examples of species that run through most of their repertoire before starting over include the following:

Marsh wren (western) (p. 60)	Varied thrush (p. 159)
Veery (p. 10)	Gray catbird (p. 49)
Gray-cheeked thrush (p. 128)	Brown thrasher (p. 92)
Swainson's thrush (p. 161)	Northern mockingbird (p. 73)
Hermit thrush (p. 88)	Green-tailed towhee (p. 146)
Wood thrush (preludes only) (p. 102)	Fox sparrow (western) (p. 145)

Immediate variety, in "packages"

A male plays with one set of songs for a while, then offers another set, eventually returning to the first package. Or possibly he creates an entirely new combination of songs, a novel package.

Yellow-breasted chat

Aptly named, he is brilliantly yellow-breasted, and he "chats," with what seems an endless variety of pops and gurgles, rattles and caws, squawks and chuckles, mews and hoots, blurting it all out in a halting, unhurried fashion. But listen more closely and the variety of sounds is not endless. In this example (♫433), he plays with five different songs for some time, two of them, the *BOB* and *WHITE* notes (heard first at 0:10), blatantly mimicked from the northern bobwhite (p. 44). Eventually, he abandons all five of those songs and introduces an entirely new package of songs. Can you hear when he does that? You will know for sure when you have heard the last *BOB WHITE*. Best of all, the chat often sings in the dead of night, on a stage all his own (one example from the West, Hells Canyon, Oregon: ♫434).

Explore 44: Listening for chat song packages.

Do the chat's "song packages" have a stable composition? For the bird in ♫433, for example, are the *BOB* and *WHITE* notes always in the same package, together with the other three sounds? It might take hours of listening to a single chat to find out, as you would have to wait for him to return to the *BOB* and *WHITE* notes after he delivered all manner of other sounds. All it would take is some time and fine listening to learn how the singing chat stores and organizes the songs in his brain. As of now, it's uncharted territory.

Yellow-throated vireo

The yellow-throated vireo offers a song about every two seconds, each only a third of a second long, with silent intervals between songs about six times the length of each song. They are much like red-eyed vireo songs, but with a buzzy, harsh quality to them (much like the burry songs of two western species, the Cassin's and plumbeous vireos). In the

first example (♫435), sonagrams in Raven Lite show that the first five songs are all different (A B C D E), but then he settles into a pattern, revealing a package of two songs as he alternates D and E for some time. Eventually, at 1:52, he returns to the first package again (A B C) and adds F, completing the package of four songs. In this section he sings A B C F A B C F A B C F A B C F A B C F A. And how about the next day (♫436)? It's the "D E" package again! Same bird, with the same package of two songs.

Explore 45: Song packages of yellow-throated vireos.

(See website for details.)

American robin

Instead of registering just a stream of those low caroled notes, concentrate on the intricacies of each. Before long, you will likely hear at least one unique carol that stands out from the others (but beware, as some robins are far tougher than others in this regard; move on to another robin if you have difficulty). Hear how he can play with that one for a while, often alternating it with several other carols, and then

it disappears for some time. A minute may go by as he sings other carols from his repertoire, but eventually the package of songs with your unique carol will return. It's hard to miss the unique carol at 0:04 in ♫437, for example; you hear it four times in the first 15 seconds, then not again until 0:37.

You can't miss another unique carol in ♫438. Brace yourself and be ready at 0:06. Smile. It's special. Over nearly half an hour, this robin offers an easy entry into understanding just how a robin pieces together his singing program.

Every singing robin provides a good listen, because the program can vary so much from one time to the next, from bird to bird, and through the season. There is no such thing as "just a robin."

More about robins: "Birds Sing and Call" (p. 6).

Immediate variety, recombining song elements

When the brief sound units within a song are rearranged in many different combinations, as with Townsend's solitaires (p. 82) and sage thrashers (p. 83), a listener cannot know what is coming next. Especially if a bird has a large number of song elements (e.g., more than 700 in a sage thrasher), the variety of songs that a male can sing is essentially limitless, as no two songs are likely to be the same. Here's a favorite example of a bird that sings a greater variety of songs by recombining the song elements in its repertoire.

Wood thrush

Let's explore what makes wood thrushes so special. Dissect each song and we hear a couple of soft introductory *bup bup* notes, then a beautiful series of fast-paced, pure tone notes (a "prelude") followed by a somewhat percussive trill (the "flourish"). Try listening to these two sessions from the same male, minutes apart: ♫439, ♫440.

Dissect the songs of this male and we learn that he has 7 different preludes (A–G) and 11 different flourishes (1–11), and that he recombines the preludes and flourishes to produce a considerable variety of songs. Listen to the beauty in just his 7 preludes, at normal, half, and quarter speed (♫441), then to the 11 flourishes, at full, half, quarter, and eighth speed (♫442; I've included one-eighth speed because the flourishes are so much more intricate than the preludes).

One special feature of a wood thrush performance is that successive songs are always different: They never have either the same prelude or the same flourish, thus creating a wonderfully diverse sequence of songs. Imagine a wood thrush singing one particular song many times before switching to another, as so many songbirds do (p. 94); most of the magic is lost, as revealed in ♫443, in which I have simply repeated one of his songs ten times before switching to another. Even when the flourishes all differ but the prelude stays the same for two or three songs (♫444), it is no longer the wood thrush that we cherish.

More about wood thrushes: "Not One But Two Voice Boxes" (p. 38) and "How Birds Go to Roost and Awake" (p. 108).

Explore 46: Listening for patterns in how songbirds present their song repertoires.

By now, your ears are well tuned for fine listening. With every singing bird, you will subconsciously listen to the pattern in each song, comparing it to the next, and the next, wondering exactly how the bird is moving through whatever songs he knows. You will be listening for a unique song (a "handle") that you are confident you will recognize if you hear it again, and you will pause long enough to get a sense of at least the short-term use of that particular song. You will linger more often, wondering when such a unique song might recur, perhaps even staying with a singing bird for hours on end. Without even thinking, you will be asking questions, and answering many of them, more and more effortlessly exploring the world of birdsong. It's a good life!

Fifty years ago I knew a graduate student, coincidentally named Robin, who was going to "do" reptiles because everything was known about birds. I have friends who are "doing" butterflies or dragonflies because they have "done" birds. Now, there's nothing wrong with reptiles, butterflies, or dragonflies, of course, but the birds I know offer limitless opportunities, with so little known about them!

WHAT? "SONGBIRDS" WITH NO SONG?

How odd. Here are a few examples of those puzzling songbirds who have abandoned their rich song-learning heritage, and instead simply have nothing that we would recognize as a song. Over evolutionary time, for unknown reasons, they have "lost" their song. Never does one encounter a male (or female) of one of these species broadcasting something loud and long and complex from the treetops, or anywhere else for that matter.

Cedar waxwing

From the gregarious cedar waxwings, one hears two basic types of calls: a high, thin *seee* and a more buzzy *bzee,* which appears to be generated by rapidly alternating sounds from the two voice boxes. Hungry young birds just out of the nest, and a female demanding food from her mate, all use the *bzee* note (♫445). Just before flying, as if coordinating their departure, waxwings in a flock usually shift from their *bzee* to an intense *seee* (♫446). It may be more appropriate to think of these two calls as the extremes in the waxwing's vocabulary, and all manner of intermediates occur in between, depending on what the waxwing is attempting to communicate in the moment.

Evening grosbeak

The evening grosbeak is a relative of the singing house finches (p. 31), purple finches (p. 161), and goldfinches (p. 134), but has no obvious song. The first clue that these grosbeaks are near during the nonbreeding season is often a chorus of loud, piercing, down-slurred pure-toned notes, maybe heard as *cheer,* or *peeer,* the actual characteristics of the note depending on where, from the Atlantic to the Pacific, one is listening (0:03 in ♫447). A second common call, heard when the birds are closer, is a quieter, raspy trill (e.g., at 0:01). How grosbeaks function with what seems such a limited vocabulary is unknown, but they do just fine, of course.

American crow

Crows *caw,* of course, loudly, and from the treetops, but all the noise crows make never feels like a song to us. In ♪448, two crows call from a treetop, apparently communicating with several distant crows, and you can hear a small sample of how the *caw* can be changed in frequency, duration, intensity, rhythm, and

tonal quality. Hear also how the two crows can match the nuanced pronunciation of each other's calls, matching each other and even completing each other's "sentence" with exactly the same *caw* (more on this kind of matching, p. 58).

Crows belong to the family Corvidae (p. 174), and it seems that all members of this family have a large repertoire of calls, but nothing like what we typically think of as a song (see also blue jay, p. 78; Steller's jay, p. 77; common raven, p. 59). Despite the great variety of caws available, a single crow can stay "on message," giving the same caw for several minutes (e.g., ♪449). Listening to a flock of excited crows mobbing a predator is nothing short of hair-raising; the flock in ♪450 is dive-bombing a dead magpie lying on the ground. What? Yes. Go figure.

Pinyon jay

The pinyon jay is another corvid with an abundance of different calls, but never a sound that would seem to function as a song. What a fascinating life history they have, as they depend largely on harvesting, caching, and retrieving pine seeds; they are highly social, nesting in large colonies, after which adults and young leave the nesting colony and move about their large home range. Listen in ♪451 to a large flock of hungry young birds, all of them demanding

to be fed, with calls of the adults interspersed throughout (e.g., at 0:01, 0:12, 0:29, 0:35, 0:48).

With the crows and jays, a small clarification is necessary: Although they never rise to the treetops and deliver something we might think of as a song, in private, in close quarters, either all alone or with close companions, one can on rare occasion (in half a century of listening,

I have never heard it myself) hear a rambling kind of twittering or musing that one might think of as, well, a "song." It's somewhat like the plastic, rudimentary songs of a songbird who is still working on getting it right, as if the corvids have stalled their song development in the early developmental stages. So a blue jay, for example, perhaps secluded by itself in a dense evergreen, may ramble quietly for a couple of minutes, sounding like a muted gray catbird or an American robin; a close listen almost suggests that he is gently and peacefully rehearsing all of his raucous calls that he'll deliver in a more public venue, but the delicious variety of whistles, whines, clicks, twitters, buzzes, chortles, mews, and more is puzzling.

An example from the California scrub-jay illustrates this rambling song well (♪452); furthermore, not only was this jay bird hidden in the bushes, but it also seemed to "sing" mainly when the wind gusts were loudest, or when a vehicle passed, as if it were having a private conversation with itself and really did not want to be heard. As you explore the world of birdsong, be on the listen for this type of corvid song, but don't bet your life on hearing it!

Chestnut-backed chickadee

The chestnut-back is no doubt the most handsome chickadee, but what a puzzle, as he has lost his whistled song. Unlike the black-capped (p. 8) and Carolina chickadees (p. 87), and their close relative the mountain chickadee, all three of which have an obvious song and a variety of calls, the chestnut-backed seems to have only the calls. This tiniest of chickadees ekes out a high, thin, scratchy *tsick-i-see-see,* and like other chickadees has a variety of other calls as well. During a dawn chorus that I experienced, when the great horned owls were still "singing," and while other songbirds were singing lustily, the chestnut-backed chickadee offered only what sounded like his *tsick-i-see-see* call (♪453), which can be heard in a variety of other situations as well (e.g., ♪454, ♪455).

Bushtit

The bushtit is an odd bird in North America, the only representative of "long-tailed tits" that are primarily an Old World group. And how rare, it seems, to have a lone bushtit sputtering all by itself (♫456), *tsit tsit . . . tsit . . .* probably agitated for some reason at me. Vary this fundamental *tsit* call in several dimensions, such as loudness, duration, frequency, and overall emphasis, and you have the basic sounds that are most often heard from these minuscule

birds. Put a dozen or more bushtits together, as they are so often seen in flocks and in group huddles roosting at night, and the combined effect is a pleasant group twitter, the birds seeming to banter constantly with each other. Do they in addition have a "song"? A rarely heard, more complex vocalization that might qualify? Maybe. Some (human) listeners swear by it.

Explore 47: What do songless songbirds do during the dawn chorus?

For these songbirds without a song, the dawn chorus (see "Energized Dawn Singing," p. 112) is worth exploring more intently. That time of day is most likely to tell us what a songless bird considers to be its song, if it has one, or which of its sounds functions in the same way that the songs of other songbirds do. When all other species are singing intensely during the dawn chorus, just what are these supposedly songless songbirds doing? Good question, but one with no obvious answers from any ornithologists, it seems. Are these songless songbirds passionless during the dawn chorus, or do they express their passion in some way, perhaps with vocalizations that we consider to be "calls"?

7. When to Sing, and How

HOW BIRDS GO TO ROOST AND AWAKE

Many birds have a predictable ritual when they go to roost in the evening and leave the roost in the morning. You just have to make a small effort to discover what it is.

Going to roost

Pileated woodpecker

Toward evening, around sunset, listen for his or her wild cackling, an untamed, irregular series of primeval *kuk* notes lasting a few seconds, *kuk-kuk-kukuk-kukuk-kuk-kuk-kukuk* . . . It may well be a pileated heading to its roosting tree. Continue listening, but walk stealthily toward the calling bird, waiting for another series, and perhaps another, homing in on the bird until you hear and maybe see the finale, the intense, "laughing" cackle of a pileated woodpecker

in flight as it swoops through the forest to the tree cavity where it will spend the night (♫457). If you lose the bird one evening, return the next and pick up where you left off, as the roosting cavity will likely be the same.

More about the pileated woodpecker: "Mechanical (Nonvocal) Sounds" (p. 21).

Wood thrush

Some minutes after sunset, throughout the singing season and until the birds have migrated south from our winter, listen first for their gentle *tut tut* notes, announcing that the evening routine is about to begin. Perhaps another bird chimes in, either the mate or birds on a nearby territory, the soft *tut* soon giving way to sharper, emphatic *whit whit* calls, the favorites of mimicking white-eyed vireos (p. 77).

Back and forth the birds may call, fading to *tut*s, surging to *whit*s, perhaps even an outbreak of song or bits of song, until finally the energy fades and the forest transitions to nighttime quiet, the birds having gone to roost (♫458).

More about wood thrushes: "Not One But Two Voice Boxes" (p. 38) and "How a Repertoire Is Delivered" (p. 94).

Awaking

Great-tailed grackle

At any time of the year, one is treated to a marvelous cacophony of sound at a roost of these grackles. During the nesting season, most of the breeding males are at least three years old and stay on their own small territories, away from the roost. That leaves the one- and two-year-old males and nonbreeding females in the local communal roost. Just listen to the enormous racket of the waking colony, with all the young males singing (♫459). Many songs begin with what sounds like ruffled wings

or crackling twigs or thrashing branches, followed by ear-shattering squawks and yodels and screeches, some of the most remarkable and diverse sounds heard among birds.

More about great-tailed grackles: "Music to My Ears—Author's Choice" (p. 164).

Blue jay

During winter, each blue jay roosts where it pleases, but they all listen to each other when they awake in the morning. The jay in ♫460 roosts in a 30-foot hemlock overnight, then begins calling in the morning, *jay . . . jay . . .* in one of the more pleasing jay voices. Nearby, two other jays call more harshly, and our jay, after a listening pause, switches (at 0:40) to match their harsh call. When the background jays switch to the less harsh version (at 1:04), our jay again follows (at 1:17), returning to the call it had begun with. Songbirds without an obvious song, these jays play the same matching games with calls that other songbirds do with their songs (see "Song (and Call) Matching," p. 58).

More about blue jays: "Mimicry" (p. 73).

Swainson's thrush

How fascinating to be waiting among these thrushes an hour before sunrise, listening for them to awake, and then for a few minutes to hear only calls, not songs. With headphones on, listen in ♫461 to the variety of calls from the Swainson's thrush in the foreground, to both subtle and striking differences in intonation. Then listen more deeply, and you hear a thrush in the background using the same calls. When one bird switches to a new call, others typically switch as well in this calling chorus, until eventually they break into song.

More about the Swainson's thrush: "Music to Our Ears" (p. 153).

Veery

With water dripping from the trees after overnight rains, the veeries in ♫462 also awake with calls, not songs. Three birds are clearly audible, all using the same call. At 0:50, they all switch to a new call. How many different calls these veeries use, or whether the variety of intonations is infinite with birds making up new ones on the fly, no one knows.

More about veeries: "Birds Sing and Call" (p. 6).

More about veeries: "Birds Sing and Call" (p. 6).

> ### Explore 48: Dawn-calling choruses in Swainson's thrushes and veeries.
>
> What is it about these thrushes, and no other species I know, that leads them to call like this rather than sing? Does the intensity and duration of calling vary with the season, from the time they arrive from migration, through pairing, nesting, and into late June and July? How many birds within a territory are calling? Initially just one bird, the male who has arrived first from migration, then two birds after pairing? Or how about after the babies have fledged? Does the entire family awake in a calling bout? Do they also call like this upon awaking in South America during the nonbreeding season? Or in migration? So many questions, so few answers.

Northern mockingbird

A mockingbird awaking on a winter roost sounds as if it is all business (♫463). One of its primary tasks for the day is to defend its food supply, such as palm fruits or winter berries, until insects become more abundant again in spring. The defense begins in earnest well before sunrise when, to our ears, the mockingbird uses a variety of harsh, angry-sounding calls, growls that sound like *CHRRR,* or *CHIT*s and *CHAT*s. Mockingbird country erupts with these calls well before sunrise, each bird seemingly eager to announce its presence and its readiness to defend what is his or hers. The going-to-roost calling in the evening is much the same, with the last calls fading with last light.

More about mockingbirds: "Song (and Call) Matching" (p. 58), "Mimicry" (p. 73), and "Night Singing" (p. 123).

> ### Explore 49: Mockingbird calls at dusk and dawn.
>
> (See website for details.)

Explore 50: Cardinals awaking—13 mornings to explore!

For 13 mornings during late April and early May 2018, I recorded how a male northern cardinal left the night roost and began his day (♫464–476; see also p. 19 for an example of an awaking male and female cardinal). Oh, the questions that come to mind! When does he begin relative to sunrise? Does he always begin with those sharp *chip* notes? With which of his dozen or so songs (p. 89) does he begin? Does he have favorites? What kind of a warm-up is involved before he sings with full gusto? How often does he match the songs of birds in the background? (Those singers could be neighboring males or his female partner; it was too dark to know.) If matching occurs, who matches whom—that is, who leads and who follows in the matching game? How many different songs does he offer over these 13 mornings? Do passing cars influence his singing? What else can you hear and learn from this male cardinal? These 13 mornings are a bonanza, but then I wonder how any patterns might have changed as nesting got under-way, during late May and into June, through July.

ENERGIZED DAWN SINGING

The pre-sunrise chorus of birdsong during the breeding season is intense, prolonged, and exhilarating. It seems that every male individual of every species is on a mission, delivering the strongest possible message about . . . something, though exactly what that something is remains uncertain (in part because professional ornithologists do not like to arise early either, one of them providing the excuse that he didn't study birdsong before sunrise because he couldn't see the birds then). One possibility is that birds sing most then because they have just awakened and each male wants to inform everyone that he is still alive. Or it is too dark to find food then, so why not sing? Or it is calm then, and sound travels better then than later in the day.

Here, I think is the best educated guess as to what the energized dawn singing is all about: While the females are listening and making their mating decisions, the males are singing in some fashion that is intended to impress the females—perhaps sparring with one another, both vocally and sometimes physically, and thereby demonstrating in some way their relative attractiveness to a female.

Because birdsong at dawn is so rich, in addition to the species featured in this section, I include examples for a variety of species throughout the book:

Mourning dove (p. 45)
Yellow-bellied sapsucker (p. 24)
Kingbirds (p. 149)
Olive-sided flycatcher (p. 84)
Wood-pewees (p. 148)
Willow flycatcher (p. 45)
Least flycatcher (p.79)
Empidonax flycatchers (p. 151)
Black phoebe (p. 84)
Eastern phoebe (p.46)
Say's phoebe (p. 84)
Bell's vireo (p. 164)
Blue jay (p. 110)
Tree swallow (p. 132)
Purple martin (p. 126)
Carolina chickadee (p. 87)
Black-capped chickadee (p. 155)
Chestnut-backed chickadee (p. 106)
Rock wren (p. 90)
Marsh wren (p. 60)

Bewick's wren (p. 167)
Mountain bluebird (p. 133)
Veery (p. 110)
Swainson's thrush (p. 110)
American robin (p. 10)
Brown thrasher (p. 92)
California thrasher (p. 33)
Sage thrasher (p. 127)
Northern mockingbird (p. 75)
House finch (p. 31)
Eastern towhee (p. 94)
Bachman's sparrow (p. 96)
Vesper sparrow (p. 67)
Black-throated sparrow (p. 95)
Dark-eyed junco (p. 81)
Great-tailed grackle (p. 109)
Western meadowlark (p. 151)
American redstart (p. 139)
Chestnut-sided warbler (p. 139)
Tanagers (p. 147)
Northern cardinal (p. 112)

Great crested flycatcher

By day, we hear the occasional raucous outburst from the canopy (♪477), but how differently he begins his day (♪478). For half an hour or so, stopping well before sunrise, he methodically offers a modest phrase about every two seconds. Most often one hears (1) a drawn-out *wheeee-up,* then maybe (2) a higher-pitched and more raspy version of the *wheeee-up,* followed by (3) a faint, low buzzy note that can't even be heard unless one is close (in ♪478,

listen carefully at 0:09, 0:16, 0:21, 0:26, etc.). These three recognizably different sounds are typically given in succession, though careful analysis shows that the *wheeee-up* can vary in many dimensions, including duration, emphasis, amplitude, tonality, and more, giving him a wide range of expression.

Eastern bluebird

It is difficult for diehard bluebird lovers to admit that the eastern bluebird is not an especially great singer compared to some other thrushes (♫479). Appropriate adjectives for its daytime songs include sweet and melodious and mellow, pleasing and cheerful, charming even—all euphemisms for "okay," but not all that special. The pace is gentle, the songs simple and short, fairly quiet, consisting of just a few warbled phrases over less than a second.

But in the darkness of the predawn hour, the male bluebird transforms into an aggressive singer. His songs now sound far louder and are often delivered at breakneck speed, the time between songs filled with what sounds like a harsh, belligerent chatter as he interacts with other males in the neighborhood (♫480).

Horned lark

The horned lark is best known by his daytime song, a relatively simple second and a half of a few stuttered notes that rise to a more rapid, slurred flourish. Offered a few times each minute, it is officially called by ornithologists the "Intermittent Song" (♫481, ♫482).

But hear him an hour or more before sunrise and he is a different bird, now singing his "Recitative Song" (two examples: ♫483; in ♫484, at about 0:10, the horned lark's song emerges from that of the foreground lark bunting). The flourishes are still there, about five each minute, but he fills *all* air time between them with four to five notes each second, jumping high and low in an irregular cadence. With about 10 different flourishes and 200 different notes available to sing in unpredictable sequences between the flourishes, it is an especially invigorating performance.

Explore 51: Exploring horned lark singing with Raven Lite.

In the daytime singing of ♫482, would you believe that Raven Lite son-agrams show that the first (at 0:06) and tenth (1:00) flourishes are iden-tical, and that the second (0:13) and eighteenth (2:01) are identical? In the dawn singing of ♫483, the first (0:22) and third (0:34) flourishes are the same, as are the second (0:34) and seventh (1:11). With some sleuthing on Raven Lite, you could learn a lot more about how many flourishes he has and how he delivers them during his two types of singing.

Black-headed grosbeak

The black-headed grosbeak excels in song, just as does his eastern cousin the rose-breasted grosbeak (p. 32). Brilliant and bold, rich and rollicking, exquisite—like a robin who has taken voice les-sons, it is said. Song duration varies considerably, from a few to several seconds, and then he pauses, preparing for the next outpouring; that, at least, is his familiar daytime routine (two examples: ♫485, ♫486). Topping that, for his dawn effort he slows the pace but now sings *continuously,* as if wanting to articulate each of his artistic phrases with the greatest of care (♫487). The effect can be breathtaking (♫488).

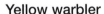

More about rose-breasted grosbeaks: "Female Song and Duets" (p. 15) and "Courtship Songs" (p. 30).

Yellow warbler

Songs of the yellow warbler sweep rapidly up and down, with high, clear, whistled phrases, often end-ing on a sharp, high note: *sweet sweet sweet I'm so SWEET.* During the day, a male often repeats one of his songs over and over (♫489) and reserves about a dozen others for use during the dawn chorus and during aggressive encounters with other males (♫490). Though all of his songs are of the same sweet quality, before sunrise he leaps from one song to the next. Successive songs are almost always different, and he also chips frantically between songs. Here, for the Idaho male in ♩490, is the song program for the first three minutes, with the first occurrence of each song type underlined: (1) A B C D E D F B B G H I E D; (2) F J C A KL K K B I E D F D; (3) F J C A M A L N B G O I E D F. That's 15 different songs! What a remarkably different performance than in relaxed singing during the daytime. If by ear you want to gauge the intensity of singing in a yellow warbler, listen intently to each song, and simply declare the next song "same" or "different." In dawn singing, you will record mostly "different," but during the day "same."

> ### Explore 52: Warbler dawn choruses are special.
>
> Warblers provide extraordinary listening opportunities. Many of them have specialized dawn and day songs, much like the yellow warbler (e.g., American redstart, p. 139; chestnut-sided warbler, p. 139; black-throated blue warbler, p. 28). Others seem not to care about such matters; they use the same songs during the dawn chorus as during any other circumstances throughout the day (common yellowthroat, p. 11; Tennessee warbler, p. 53; magnolia warbler, p. 54; ovenbird, p. 138). Pick your favorite warblers and go listen! What does a common yellowthroat, for example, do during the dawn chorus to show some passion that is not revealed during daytime singing? Are songs longer, or shorter, or delivered at a faster or slower or more continuous pace? Does a male engage in heated exchanges with neighboring territory owners? Listen to just a few warblers at dawn and you will be astonished at the variety of expression among them. Because of the early hour, most likely you will listen alone, revealing the reason we know so little about these warblers at dawn: Professional ornithologists like to sleep in too!

Orchard oriole

We hear the orchard oriole's daytime song from the treetops, a wonderful, rollicking jumble of buzzes and piping whistles. Songs are fast-paced, with great variety, as the songster plays with the overall duration, and also with repetition and rearrangement of phrases within the song (♩491).

In the predawn darkness, however, he often sings in a small bush low to the ground (♫492), doubling his pace. If careful, you can walk up to the bush until you feel you could reach out and touch him, though it is still far too dark to see. What boundless energy and enthusiasm one now feels from him, as he is *rocking!* He is on a singing rampage, maybe all by himself, but on occasion matching wits with

another male in a nearby bush, as male songbirds often engage each other in these kinds of heated, predawn exchanges at their territorial boundaries. Here, in fact, is the immediate neighbor of the male in ♫492, singing about ten yards away: ♫493.

Red-winged blackbird

The most common of birds can provide the most uncommon and thrilling experiences. Stand with me in the dark, an hour before sunrise, beside a local marsh in northern Michigan, waiting for the male red-winged blackbird to awake. Recall first how he sings routinely during the day, offering one of his half-dozen or so songs many times in a row before switching to another (♫494). Now, in the dark on

this early May morning, 38 minutes before sunrise, at 5:44 A.M., he calls, and half a minute later offers his first song. What extraordinary energy he devotes to his effort through the next three-quarters of an hour, roughly until the female arrives at 47:12 (♫328), about ten minutes after sunrise.

I eagerly explored his first 50 minutes of activity on my own. First, I inspected sonagrams of his songs in Raven software and documented that he sang seven different songs (A–G). Then I listed his songs over the ten 5-minute blocks of time. Wow, over the first 15 minutes, back-to-back songs are almost always different; they are the same only once (F F in minute 9—underlined on p. 118)! Gradually, as he winds down, he sings strings of the same song, more like his relaxed daytime singing.

Sequence of song types during the 50 minutes

Each row comprises five minutes, with a period separating the minutes; underlined letters represent strings of the same song.

A.G F. D A. B C G.R D A
B G C. F A. G B C. <u>F F</u> C E B.G E C B
C F G.C E B A.G F B F. D A G E.C D G A
B <u>E E</u> <u>G G</u>.B A D B.A <u>G G</u> E.C G B A. F C D G
E C G A.B A D <u>E E</u>.<u>D D</u> B <u>G.G</u> <u>F F</u> . .
<u>F</u> C B.<u>G G G</u> C.<u>G G</u> <u>F F</u>. F F F.F <u>C C C</u>
<u>G G G G</u> C.<u>C C C</u>. C C.C.<u>E E E E E E</u> E
<u>E E</u> E <u>D D D</u>.D D D.D D <u>A A A A A A</u>.A A A <u>C.C C C C C C</u>
<u>C C</u> D D D D.<u>B B B B B</u> B.<u>A A A.A</u> A.<u>F F F F</u>
. <u>G G G.G G G</u> G.<u>E E E E</u> F . .

I wanted to see this in graphic form, so I used Microsoft Excel to plot the probability of the next song being different for each of those 5-minute blocks. I can now see how for 25 minutes the next song is almost always different (at least 9 out of 10 times, i.e., 90 percent), and then he transitions to longer strings of the same song well before sunrise.

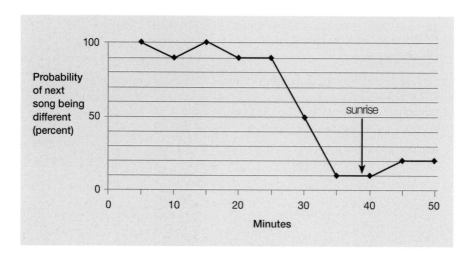

I tried another graph, showing how at first he leaps from one song pattern to the next, then gradually settles on singing longer strings of one particular song.

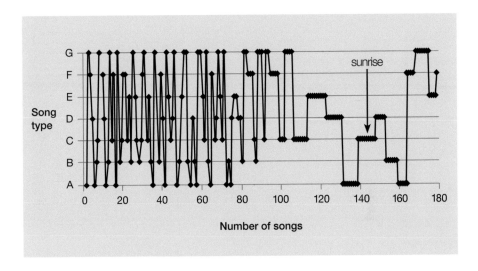

And I'm hardly finished. I want to know so much more, especially about how he uses his calls throughout this dawn chorus, and how he behaves after the female arrives at 47:12. But I need to leave something for you to explore.

More about red-winged blackbirds: "Female Song and Duets" (p. 15) and "Song (and Call) Dialects" (p. 63).

Explore 53: Individuality among red-winged blackbirds.

How befuddling! As I was entranced by the male red-winged black-bird in this account, I increasingly realized that his neighbor barely sang throughout the entire dawn period. He seemed, in a word, passionless! He was also paired, and he and his mate foraged together on their territory, but without all the ruckus of their immediate neighbors. The striking differences are mysterious, but perhaps could be explained if one knew more about the two pairs. A simple description of how several male red-wings in a marsh behave through the dawn period would be enlightening, especially through the entire breeding season. That description would inevitably lead to better ideas as to why the males differ so much in their behavior, and why they sing and call the way they do during the predawn hour. Be aware: You may find no obvious reasons as to why the males differ, and may in the end just chalk up the behavioral differences to their individual personalities.

Explore 54: An *extreme* blackbird challenge for you!

(See website for details.)

Spizella sparrows

Check your field guides and you will find four closely related species in the genus *Spizella* (p. 177). Let's explore how each of them sings during the dawn and day, realizing that they evolved their current behaviors from a common ancestor. Similarities among the species would suggest behaviors retained from their common ancestor; differences have arisen since then over evolutionary time. Let's "compare and contrast" (three of my favorite words from essay exams since kindergarten, it seems) the day and dawn singing of these four species.

For additional comparisons among closely related species, see "Song Changes over Evolutionary Time, from Species to Species," p. 147.

Chipping sparrow

From the treetops a chipping sparrow sings his longish daytime song, up to five or more seconds in duration, the same song over and over, the only one he knows (♫497; see also p. 56). But during the dawn chorus he descends *to the ground,* using his same song there, but now it is highly fragmented as he spits out just a few phrases at a time from his longer daytime song. He spars there with other males who have left their own territories and gathered in a small dawn singing arena.

In ♫498, you can hear in the first four seconds how three males spit song fragments at each other, each one recognizable by his unique song (see p. 137). Listen to how one of the three birds disappears after 0:37, and how a loud fourth bird has joined them by 0:43. For half an hour or so they sing and fly animatedly about, the composition of the group often changing as individuals seem to come and go, probably joining different singing arenas in the broader neighborhood. Eventually, one bird may rise into a nearby tree, singing longer songs there, or the birds may disperse to their own territories, announcing an end to the ground game.

More about chipping sparrows: "Song (and Call) Dialects" (p. 63).

Field sparrow

The daytime song of the field sparrow (♫499) is the familiar "bouncing ball" of clear, musical whistles, a series of rather plaintive whistles slurred and gradually, sometimes abruptly, accelerating into a trill (i.e., overall, an accelerando, p. 154). Like chipping sparrows, each male has only one song that can be heard throughout the day.

Now hear how entirely different his song is during

the dawn chorus (♫500); the tonal quality is the same, but his song is longer and more elaborate, consisting of four or more distinct parts. He also chips excitedly between songs, sometimes seeming to forget to sing (listen to 21 seconds of continuous chipping beginning at 3:04). Field sparrows have entirely different dawn and day songs and call intensely between dawn songs. No chipping sparrow behaves like this!

More about field sparrows: "The Music in Birdsong—Accelerando" (p. 154).

Brewer's sparrow

It is just one simple, unremarkable, two-parted song that each male Brewer's sparrow sings during the day, perhaps heard as *zreeeee zrrr-zrrr-zrrr-zrrr,* or from another male *tir-tir-tir-tir-dee-dee-dee* (♫501, ♫502). Now stand among them in the dark well before sunrise and you could not possibly know these singers are of the same species. They sing con-

tinuously, alternating long series of high, buzzy, insectlike *zeeet* calls and high-pitched trills with a spectacular, canary-like ramble that sweeps down the scale. Neighbors sing near one another on territorial boundaries, such as these two males in Colorado (♫503, ♫504), making it difficult to isolate just one bird in a recording. Perhaps they even leave their daytime territories to visit males elsewhere, much as chipping sparrows do. The performance is stunning! I know of no more exhilarating dawn chorus among songbirds.

For a you-are-there kind of listen, try ♫505, for the full 24 minutes: Slip the headphones on and close your eyes (it's too dark to see the birds anyway). Imagine the intense aroma of sage in the early morning as you stand in place, your parabolic microphone held steady and always aiming at the same small spot on the ground just five yards in front of you. How absolutely extraordinary! Multiple birds are singing and flitting about, no doubt posturing in much the same way that chipping sparrows do at dawn. Every day during the breeding season these sparrows carry on like this!

> ## Explore 55: The thrill of dawn-singing Brewer's sparrows.
>
> For the thrill of a lifetime (should be on every bird lover's bucket list), I encourage you to experience for yourself a live performance of dawn-singing Brewer's sparrows. Arrive with plenty of warm clothing at least 90 minutes before sunrise, set up a chair in the sagebrush, and get comfortable. About half a dozen males will soon be within earshot, all singing and calling continuously and simultaneously. In the dark, edge out of your chair and sidle closer to one bird, and just try to follow along. Collect some numbers as to how much he calls, how much he sings, how long the songs are—whatever strikes your fancy. Or forget the numbers and just listen! Don't resist the urge to return on another morning or two. To get an eyeful of what you might experience, import ♫505 into Raven Lite and watch the 24 minutes documented there. We know almost nothing about what these sparrows are doing during the dawn chorus. A simple description of how the dawn performance changes through the breeding season would be most enlightening.

Clay-colored sparrow

How the clay-colored sparrow compares to the other three *Spizella* species is unclear. Listen to a daytime singer and you will hear one buzzy, insectlike song over and over (♫506). Perhaps each male has just one song that he sings throughout the day?

During the dawn chorus, males often chip continuously between songs, but the chip calls can be so different from male to male. Each male seems to sing just one song, but how different their dawn songs can be from each other, consisting of about 1 to 4 slow buzzes, or more than 15 rapid ones (compare ♫507 and ♫508).

We don't yet know how many songs each male has, which song or songs are used during daytime and dawn, and whether the overall characteristics of the dawn songs and day songs are different. For someone living in clay-colored sparrow country, here is some fine exploring that could be done.

Explore 56: You can unravel the case of clay-colored sparrows.

(See website for details.)

Explore 57: Choose your own dawn chorus experience.

During the dawn chorus it seems that every flycatcher and every songbird (and many nonpasserines as well) expresses an unbridled passion that you can't hear at other times of the day. For some good dawn listening on your own, you could start with one of the examples I have offered previously, but that is just a beginning, as hundreds of other species await your attention. Among my favorites are—well, my list of favorites has grown to include almost every flycatcher and songbird I know: kingbirds, wood-pewees, phoebes, crested flycatchers, the dozen species in the genus *Empidonax* (p. 151) . . . That's just the flycatchers, and then there are all the songbirds! Okay, there may be a few species who seem to do nothing special at dawn (perhaps some of the songless songbirds, p. 104), but so few that they are tough to find.

NIGHT SINGING

Expected night singers

Some birds do most or all of their business during the night, such as nightjars and owls. We expect to hear them during the night.

Chuck-will's-widow

One could hardly know that a mysterious creature like the chuck-will's-widow even exists, except for its incessant singing through the night: *chuck-WILL'S-WID-ow, chuck-WILL'S-WID-ow, chuck-WILL'S-WID-ow, chuck-WILL'S-WID-ow* . . . about one every two seconds, on and on, with an emphasis on the second and third phrase of the song. He sings mostly at dawn and dusk, it is said, but then,

who is out there through the night to contest that statement? Under a full moon, and when unpaired, a male can sing thousands of songs through the night, much like other nightjars (see p. 27 for eastern whip-poor-will).

Listen in ♫509 as two males countersing at their territorial boundary. Notice the distinctive voice of each, as their songs differ in duration by about a tenth of a second (see "Each Individual Has Its Own Song," p. 137). In a second example (♫510), after about a minute the male approaches the microphone more closely, then hurries his last song (1:57) as he is visited by a second bird, most likely a female; what must be his low, guttural *chuck* calls are then overlapped by her longer, half-second growls.

> ## Explore 58: Night singers!
>
> (See website for details.)

Barred owl

Who-cooks-for-you, who-cooks-for-you-all. Such is the well-known mnemonic of the barred owl's common call (song?). The female calls, too, and even though she may be a third larger than the male, his voice is still the lower one, perhaps because of a larger, testosterone-induced syrinx. These owls make other sounds as well, such as an ascending series of six to nine *who* notes with the familiar ending, *who-who-who-who-who-who-who-who-all,* or an isolated *yoooouuuuuuu-all,* the first note now drawn out. And there are no more hair-raising sounds in the night than the maniacal caterwauling of two barred owls (♫511); it is an otherworldly shrieking and raucous hooting, a minute or two of cackling and cawing and gurgling, sounds one might expect from two dueling demons. Throughout, her voice is always higher, but her voice is also distinguishable by having more vibrato in the *all* note, more of an *allllllllll* than the male's briefer *all*.

> ### Explore 59: Owling.
>
> Nearly 20 species of owls occur in North America, offering plenty of nighttime listening for those so inclined. Owls who offer great listening opportunities are the ubiquitous great horned owl (♫370, ♫453, ♫577) and the two screech-owls, eastern and western. A favorite of many owlers, for their hair-raising, demoniacal caterwauling, is the barred owl. Here's how you could get started: Around Christmas every year, countless birders join the Christmas Bird Count throughout North America, now extending into South America. Find the count organizers in your area and join those "owlers" who will be out in the night, prowling about for whatever owls they can find.

Surprising night singers

Some night singers are a surprise to us, because we think of them as normally active during the day. Here are a few examples.

Limpkin

The limpkin is so curious a bird, like an oversized rail, but unique and in a family all by itself with no close relatives. They are active by day in their south Florida haunts, foraging in shallow water for their favorite food, apple snails. One *can* hear them throughout the day, from dawn to dusk, but it is a nighttime vigil with them that will be never be forgotten (♫512). Shattering the

muted sounds of frogs and insects, his voice is now a deafeningly loud scream, amplified by an elongated, folded trachea and enlarged bronchial chamber. It's a piercing cry that rises and then falls, *kur-r-ee-ow kur-r-ee-ow kur-r-ee-ow kur-r-ee-ow kur-r-ee-ow kur-r-ee-ow,* continuing for half a minute or more. Up close, one hears a brief stutter at the beginning of each scream, and a low-pitched, soft stutter at the end.

Sometimes he clucks a while before the first scream, and sometimes the screams are a simpler *keoow.* Never to be forgotten!

Great blue heron

By day they are mostly silent as they forage elegantly in ponds and streams throughout much of North America. Camp out near an overnight roost that includes great blue herons, as I did among wading birds in the Florida Everglades, and listen to them sound off through the night. Sporadically and contagiously they call out in their hoarse trumpeting voice, individuals joining from near and far, *fraaahnk, fraaahnk, fraak, fra.* In ♫513 are concentrated a few of the calls heard throughout the night,

and then, around sunrise, hear the near bird take flight (at 2:06), his calls fading as he flies into the distance.

Purple martin

These large, handsome swallows now nest almost exclusively in human-made birdhouses, erected in backyards everywhere throughout their range, from farm to city. The older males are a uniform bluish black, iridescent in good light, the older females more a smudgy gray with only a little blue on the back; first-year birds are drabber versions of the adults. Visit any one of their colonies and just listen to the hubbub of martins swirling about and sitting on the condo units (♫514). Their calls are low, rich, and throaty, with much gurgling, chattering, chirruping, and twittering. The male courts a female with a rich and rapid jumble of gurgled notes ending in a harsh, grating croak; she chortles in response.

Early in the season, older males seem determined to attract late-migrating yearlings to their colony. To do so, they sing in flight, a few hundred yards over the colony, for two to three hours *before* sunrise (♫515). Yearling males added to the colony will help attract more yearling females, and even though a yearling female might pair with a yearling male, an older male has a good chance of impressing that yearling female and fathering some of her offspring (see p. 26).

> **Explore 60: Purple martin song flights.**
> (See website for details.)

Northern mockingbird

"Didn't sleep a wink last night; the cursed bird sang all night long just outside my window!" That is a common complaint about these mockingbirds, for they can sing through the night, especially nights that are well lit by the moon or by streetlights. Compared to the frenzied minutes of dawn singing, when he might jump to a new

song 30 to 50 times each minute, the nocturnal pace is slower, about half that, all the more stunning, though, because no other creatures are stirring. The persistent night singers are believed to be unpaired males seeking a mate, either a female just drifting through or one already paired to another male within earshot (which is a much greater distance in the calm of night than it is during the day). As he sings, he announces to paired females nearby that he is an alternative for anyone who is in a relationship with another male but who would welcome a change.

Listen to the slow pace of a predawn-singing western mockingbird, recorded before the dawn chorus of other birds has begun (♫516; hear the cactus wren imitation at 0:33). An unpaired eastern mockingbird sings with the same slow pace through much of the night, but mimicking only eastern birds, of course (♫517). Compare the pace of this night singing with that of an eastern dawn singer, in ♫338.

More about mockingbirds: "Song (and Call) Matching" (p. 58), "Mimicry" (p. 73), and "How Birds Go to Roost and Awake" (p. 108).

> **Explore 61: Sage thrasher songs, through the night.**
> (See website for details.)

Connecticut warbler

Breeding in the spruce-tamarack bogs and moist boreal forests of south-central Canada is the little-known Connecticut warbler, all by itself in the genus *Oporornis*. Each male has but one loud, jerky song, consisting typically of a four-syllable phrase repeated two to four times, as in *chup-a-weepo chup-a-weepo chup-a-weep,* the first or last phrase often incomplete. Loud, sharp, piercing, *loud,* ringing, reso-nant, and *LOUD,* this remarkable song is appreciated best when just a few spring peepers are calling in the still of a calm night (♫532). Males may begin singing about two hours before midnight and then continue all through the night and into the day, until late afternoon, with hardly a breather, before resting up, it seems, to repeat it all over again the next night. A male who does so no doubt sings for the same reason that the nocturnal mockingbird does: He is unpaired and seek-ing a mate. Given the strength of the vocal beacon beaming into the heavens, any night-migrating female seeking a mate will be able to hear her opportunity from a mile or more away. The male's strategy seems so logical that it is a puzzle why unpaired males of more species do not use it.

Gray-cheeked thrush

Throughout those short, not-really-dark nights in Alaska, with the sun just barely dipping below the horizon about midnight, hovering there, and then reappearing just a few hours later about 3:30 A.M., what's a songbird to do? During early June, along the Denali Highway, many of the gray-cheeked thrushes sing through much of the short night, beginning about 12:30 A.M. and continuing through sun-rise. Join this thrush, and others in the background, for an hour just after midnight on June 15, a few days before the shortest night of the year (♫533). In other habitats, Swainson's thrushes were also singing through the night (see p. 161).

More about gray-cheeked thrushes: "More Music to Our Ears" (p. 160).

Ovenbird

The ovenbird is well known for his sharply enunciated, daytime *teacher* song. The male in ♫534,
recorded near my New England home, sings at his
daytime clip of one song roughly every ten seconds
(until the significant disturbance at 2:15, when he
seems to pause extra long while collecting himself).

There's more to ovenbirds (♫535): At twilight,
during both dawn and dusk, but also throughout
the night, the ovenbird can perform a dramatic aerial display. He calls
softly at first, maybe hopping up from branch to branch, accelerating
his chipping before launching into the air. Soon he is hovering or circling in labored flight above the canopy. With spread tail and quivering
wings, he sings an ecstatic volley of warbles and slurs and twitters,
rising and falling in pitch, the source all a mystery to the casual listener
until the bird slips perhaps just a single *teacher* phrase into the song.
Just as quickly, he drops back down, swallowed by the darkness of the
forest. But sometimes, far less dramatically, he offers this song while
flying in the subcanopy, or even from a perch. (See next section, "Singing and Calling in Flight," for more on this topic.)

More about ovenbirds: "Each Individual Has Its Own Song" (p.
137) and "The Music in Birdsong—Crescendo" (p. 157).

> ### Explore 62: Flight songs of ovenbirds.
>
> The flight song of ovenbirds remains largely a mystery. Oh, sure, it
> has been described by an ornithologist as occurring mostly around
> dawn and dusk, sometimes in the night, but that's about all we know.
> The best way to find out would be to document the pattern of its use
> throughout the breeding season (though I realize that's a lot of work!).
> From first arrival of males to courting a female to egg-laying to incu-
> bation to caring for the young to migrating south again, exactly when
> and how often during the nesting cycle do males engage in these flight
> songs? Documenting patterns of use would be the first step toward
> understanding the functions of this fascinating behavior.

SINGING AND CALLING IN FLIGHT

Most singing and calling occurs when birds are firmly perched on something, but some species use their time in the air to achieve a special effect. The mechanical songs of the snipe (p. 23) and the nighthawk (p. 22) can occur only when the bird is flying, of course. (I think of those sounds as "songs," because they no doubt function much like the vocal songs of songbirds.) Singing from high in the air also broadcasts a signal much farther than from lower down, especially when lower down is a grassland where no high perches are available. In addition to the species mentioned elsewhere (see list, p. 136), here are a few more examples of birds who sing on the wing.

American woodcock

A shorebird without a shore, the plump, long-billed woodcock offers some fine listening during early spring. Find him in moist thickets or old fields by listening around dusk for his nasal *peent*s, all given on the ground from his small display arena (♫536). At first, he offers a *peent* every six or seven seconds, but near the end of a warmup lasting about 20 minutes he is *peent*ing at twice that rate. By the time you have slowly worked your way closer, he takes flight, departing at first low to the ground, then climbing, the wind now whistling with each wingbeat (15 per second) through his three outer wing feathers, all slender and stiffened, specialized to vibrate and twitter as he flies. The wing twittering becomes more rhythmic and melodious as he loops and spirals overhead, and when he's about a hundred yards up, rhythmic bursts of soft wing-twittering (at 0:57) combine with loud, equally rhythmic vocal chirping, about four chirps each second, a delightful, coordinated duet from wing and bill. See all of his wing and vocal song in Raven Lite as you listen, and you will smile at all he does. After nearly a minute, he is silent, and with enough light can be seen zigging and zagging while plummeting back to his ground arena, where he *peent*s a few more times before becoming airborne again.

Explore 63: Get to know the American woodcock.

If you live in the East, find a woodcock during early spring. At dusk, perhaps you will hear him first overhead, flying over an area roughly the size of a football field. When he stops his overhead singing, wait 15 to 20 seconds until you hear where he has landed, now *peent*ing on the ground; next time he is airborne, walk closer to his ground arena, to identify the site, but stop well short of it, and stop moving before he returns! Maybe the next day, an hour before sunset (in good weather, of course, dressed warmly), sneak in to about 30 yards from this arena, and settle into a comfortable spot, well hidden perhaps behind some bushes so as not to disturb him, and wait. He'll appear soon, maybe walking into his arena, perhaps flying in. How many times does he *peent*, and how rapidly, over what time period, before his first flight? How much time and how many *peent*s before the next flight? Are his display flights at all coordinated with the display flights of any *peent*ing neighbors? It'll all be over in an hour or so, so stay with him. Study your numbers and your notes and you will have come to know a whole new woodcock. Then there's dawn! Does he behave any differently then?

Willet

Quiet and unobtrusive, camouflaged in browns and shades of gray, the willet is easily overlooked—*until* it flies. Each wing is now flashing the long white wing stripe, set off with a black border, so distinctive for this shorebird. And should you not see it, the willet's presence will be known instantly by its loud *pill-will-willet*, given about once each second as it flies about its territory (♪537). The female sings too, though her *pill-will-willet* is said to be some-

what "less musical and flatter" than that of the male. Before that bout of *pill-will-willet* calls in ♪537, the same bird seemed highly agitated as it flew about its territory, interacting with what looked like an intruder, calling sharply *klik klik klik klik* (♪538).

Red-tailed hawk

When the movie industry wants to instill a feeling of raw power, no matter what the context, it seems, no matter whether the screen depicts a mute vulture or powerful eagle, it relies on the red-tailed hawk. Dubbed into the scene is this hawk's two- to three-second shrill and hoarse asthmatic scream, slurred downward with several subtle but abrupt shifts in pitch: *kee-eeee-arrr.*

During the spring and summer, we hear a *kee-eeee-arrr* and instinctively scan the skies, usually finding the red-tail soaring there (♫539). In early spring, see the male and female courting high overhead, diving and ascending, touching, each screaming *kee-eeee-arrr.* In aerial territorial squabbles, the quality and duration of the *kee-eeee-arrr* varies with the emotions, with some sounding shrill and pure, others more like a hissy steamboat whistle.

Later in the season, perhaps when we approach its nest, a red-tail might perch overhead, calling more rapidly, as with some urgency, with what sounds like a shortened version of the *kee-eeee-arrr* (♫540). Here's another perched red-tail with the same seemingly urgent calling, this time harassed by a dive-bombing American kestrel (♫541).

Nestlings utter a softer version of this call, suggesting that it is inborn, not learned from adults.

Tree swallow

Tree swallows are among the earliest spring arrivals, possibly because of intense competition for nesting sites, which are often human-built nest boxes. Especially after his mate has begun incubating, the male is one of the earliest singers in the morning. About an hour before sunrise, he begins singing while flying in roughly elliptical patterns about 10 to 15 yards above his chosen nest site (♫542). His song is rather sweet and melodious to our ears, a con-

tinuous, liquid twitter, about two brief phrases per second, over 100 each minute. Initially one hears what seems to be an endless variety of sounds, but careful listening usually parses out only two or three different phrases offered in no set order. Part of the dawn routine often involves descending to perch near the nest site (e.g., ♫543) and singing there briefly before rising skyward again.

> **Explore 64: Dawn song flights of tree swallows.**
> (See website for details.)

Mountain bluebird

A full 80 minutes before sunrise I have Old Faithful in Yellowstone National Park all to myself. The crowds that will gather every hour or so during daylight are nowhere to be seen, and high overhead, circling about, is the local mountain bluebird (♫544). Few people know him in this singing mode, simply because of the hour. Out of curiosity, I study one of his songs and see that the 14 notes occur in the following sequence: A B C D E F G B C D E F G B.

That's intriguing, how he has a favorite sequence of notes (underlined) and plays his song out by repeating that sequence twice. It's not one of the finest thrush songs, but sweet nonetheless, bluebird lovers would declare.

Common redpoll

As with goldfinches (next) and some other species in the Cardueline subfamily, a male redpoll sings in a display flight, slowly circling about his territory. Follow him, trying to keep him in sight, and you will see him land, sing from that perch, then circle about again, perhaps singing for a minute or more at a time, often landing in the same general area before taking flight again. During mid-June along the Denali Highway in Alaska, the male in ♫545

perched in a treetop near me, sang there for about 20 seconds, and then was off again, disappearing into the distance as I did my best to follow him with my parabolic microphone.

American goldfinch

Overhead during winter, a small flock of gold-finches bounds buoyantly by in undulating flight, each bird bouncing up with a few wing flaps, gliding down briefly on folded wings. But we would never notice their trampolining passage except for the persistent calling, *per-CHIK-o-ree, per-CHIK-o-ree,* or *po-TA-to-chip,* as some prefer (♫546). Most often there are four syllables, the second accented, all pleasing musical notes.

During their breeding season, late June to early July in the East after other birds have raised their young, each male circles above his territory of goldenrods and asters, thistles and chicory. He is in full display now with slow, deep wingbeats that make him look twice his size, singing nonstop his long, canary-like songs of jumbled twitters and musical phrases, punctuated every so often by a rising, question-ing note, *sei silieeee* (♫547). When more settled, he perches beside his mate, singing shorter songs and courting her with his sexy *sei silieeee,* the small details of his and her *per-CHIK-o-ree* call notes converging on each other so they can be identified as one, as a pair (see also p. 72).

McCown's longspur

With nary a tree in sight, from the sparse vegeta-tion in the shortgrass prairie of the Great Plains, the male longspur launches skyward, flying initially downwind and up to about 50 feet overhead. See him then turn into the wind, and with wings out-stretched and angled back, tail spread, back and rump hunched and fluffed, he pours out his song, a tinkling and tumbling array of notes that can continue the entire parachute ride back to earth (♫548). Just before landing he might hurtle skyward again, twice, three times, extending the song session by repeating his routine.

Lark bunting

Perched, the male lark bunting offers relatively brief songs, about two phrases over two seconds. You must wait for him to take flight to hear the bird worthy of the status "State Bird of Colorado." With his white wing patches flashing, he rises on rapid wingbeats to about 15 feet, then rows his wings more slowly and deeply, mothlike, the spread tail with showy white tips adding extra buoyancy. And how he sings (♫549)! It's a series of contrasting phrases, slow and low followed by fast and high, sweet, then buzzy, a rich and complex, delightfully

rhythmic song with slow tempo, the elements ranging from cardinal-like slurs to sparrowlike buzzes, about eight phrases over eight seconds until he glides to a landing some distance from where he began. A half-minute later, after a couple of brief perched songs, he launches again, singing on the wing, his song and flight path stitching together the small bit of prairie he has claimed as his territory.

Explore 65: Prairie flight songs.

Nesting side by side in the shortgrass prairie of the northern Great Plains, with not a song perch in sight, the McCown's longspur and lark bunting offer wonderful opportunities for exploring what flight songs are all about. Stake out a pair or two during the daytime, and then return well before sunrise. What patterns in the song flights can be documented through the first couple of hours of a morning's activity? Do these flight singers reveal any special passion during the dawn chorus when perched singers are at their most intense? How many flights? For what duration? How do flight songs compare with perched songs? The numbers you choose to collect will tell you much about the lives of these flight singers.

Western meadowlark

From the highest perch in his grassland or sagebrush territory, the western meadowlark offers some of the finest songs anywhere, rich and well-articulated whistled phrases that, over a second and a half, accelerate and drop into a rapid, gurgling jumble. Seconds later, he repeats himself, and if you stay with him, you will hear several of that particular song before he eventually switches to another of the ten or so masterpieces in his repertoire. That's his routine; in ♫550, for example, he offers the same song nine times, the female often "rattling" in response (e.g., at 0:26), but he stops singing before introducing another of his songs.

But catch him in a rare boundary dispute, perhaps in a heated exchange with another male, and it's a different story. Excitedly, he gives a low, whistled note a few times before taking flight, the song now transformed into an ecstatic, hurried twittering (♫551; ♫552).

More about western meadowlarks: "How a Repertoire Is Delivered" (p. 94) and "Song Changes over Evolutionary Time, from Species to Species" (p. 147).

Examples of other birds that routinely call or sing in flight include:

Canada goose (p. 12)

Common nighthawk (p. 22)

Chimney swift (p. 13)

Broad-tailed hummingbird (p. 22)

Killdeer (p. 13)

Sandhill crane (p. 16)

Wilson's snipe (p. 23)

Eastern kingbird (p.149)

American crow (p. 105)

Common raven (p. 59)

Horned lark (p. 114)

Cassin's sparrow (p. 96)

Yellow-breasted chat (p. 100)

Bobolink (p. 68)

Red-winged blackbird (p. 18)

Brown-headed cowbird (p. 71)

Ovenbird (p. 129)

Explore 66: Just when do birds sing in flight?

(See website for details.)

8. How Songs Change over Space and Time

EACH INDIVIDUAL HAS ITS OWN SONG

People who know birds can routinely identify species by their songs, but the birds do far better than that, because they know and recognize each other as *individuals* by their songs alone, just as we know each other by our voices alone. Sometimes the songs of individuals are strikingly different from each other and we can easily recognize them, but with other species we need to slow the songs down to quarter speed, or even slower. Studying their voice prints (sonagrams, as in Raven Lite) almost always enables us to see individual differences that the birds no doubt hear and use to organize their social relationships with each other. Because a bird typically acquires his song during the first year of life, and because the song then remains (mostly) unchanged throughout life, we can in many species use songs that we learned in one year to recognize the same individual the next year.

Red junglefowl (barnyard chicken)

Cock a doodle doo! crows the rooster, a mnemonic that dates back to a children's nursery rhyme in the early 1600s. The barnyard rooster crows to assert himself and no doubt to impress his harem of females who nest there, and so fancied is the crow by humans that rooster crowing contests are traditional sports held in a number of countries. The crow immediately says to us "barnyard male chicken," but listen again more closely and we hear striking differences in rhythm and quality among the roosters. When listening in headphones to ♩553, close your eyes and imagine sitting in this Kentucky barnyard, reveling in the variety of their voices. If in doubt about your ability to distinguish each rooster by his voice, listen to ♫554, in which you can hear three crows from each of three roosters, back-to-back for ready comparison.

More about the barnyard chicken: "What Birds Hear" (p. 41).

Ovenbird

The ovenbird has an astonishingly loud song, an emphatic, explosive crescendo (p. 157) that builds and builds over two to four seconds until the entire forest shakes with the reverberations. Each male's single standard, daytime song is typically a two-syllable phrase repeated about eight times, the familiar mnemonic being *teacher teacher teacher TEACHER* **TEACHER TEACHER.** We can immediately say "ovenbird" upon hearing it, but when listening more closely we can often hear subtle differences among the birds. For some birds, the emphasis is on the first syllable (*TEACH-er*); for others, the second syllable (*tea-CHER*); and in still others we hear only one syllable (*TEACH*).

Listen for the rather subtle differences in *teacher* intonations in the three songs from each of these three neighboring males: ♫555, ♫556, ♫557. Our ears need some help, so we use the trusted method of slowing the songs down. In ♫558, you can hear just one song from each bird, first at normal speed, then half speed, then a few excerpted *teacher* phrases from each at one-quarter and one-eighth speed. At the slower speeds we begin to hear details readily distinguished by these birds. Back in 1959, two enterprising ornithologists showed that the birds use these song differences to recognize each other as individuals.

Being able to recognize individual ovenbirds by their songs helps one hear how the two males in ♫559 interact with each other as they creep along the forest floor. They have just returned from migration, and each seems insistent on claiming this particular bit of real estate as his own.

More about ovenbirds: "Night Singing" (p. 123) and "The Music in Birdsong—Crescendo" (p. 157).

American redstart

High-pitched, thin, sharp, yes, piercing if heard nearby, but what is perhaps most impressive is the enormous variety of songs among males. Some songs end high (e.g., *tsee tsee tsee tsee tseet*), others low (*tsee tsee tsee tsee tsee-o*); many have single-note syllables, others double notes (*teetsa teetsa teetsa teetsa eet*). During the dawn chorus, a male bounces energetically among several different songs (an aver-

age of four), but reserves another song for the daytime, presumably to use more with females. Because no two males seem to agree with each other on the five or so songs each is to sing, we can use their songs to identify them as individuals, both within a year and from year to year.

In ♫560 and ♫561, hear how the same two dawn songs are used on the same territory in successive years. In the first year, this male was a "yellowstart," indicating he was one year old; in the second year, he was a true "redstart." His two neighbors during the first year had entirely different dawn songs, and each used three different dawn songs, not two: ♫562, ♫563.

In ♫564, ♫565, and ♫566 you can hear the one song that each of these three males reserves for daytime singing, when passionate exchanges among males have ceased and a male is more likely to be interacting with a female.

More about the redstart: "Learned Songs of Songbirds, and Babbling" (p. 49).

Chestnut-sided warbler

The daytime song of the chestnut-sided warbler is the well-known *very very very very pleased-to-MEET-CHA,* a striking song building to an emphatic ending that boldly sweeps up and then down the scale. Sometimes the song ends with two upswept *MEET* notes, sometimes even three. But well before sunrise, stand in the dark among these warblers and you will hear an entirely different song, a warbled ram-

ble with no emphasized ending, with animated *chip* and *chug* notes between the songs. Curiously, each male is capable of singing several versions of both his day song and dawn song, and even the same songs as his neighbors, but he chooses just one favorite song for each context, and almost always it is a song that is different from the songs his neighbors have chosen. There are only a few versions of the day song, but a seemingly infinite variety of dawn songs, so the favored dawn song can be used to identify individuals.

Try listening to the dawn singing of these four males: ♫567, ♫568, ♫569, ♫570. In ♫567, the male calls excitedly between songs, typical of behavior during the intensity of the dawn chorus. In ♫570, you will first hear seven dawn songs, and then the male transitions to his day song with the emphatic ending, *very very very very pleased-to-MEET-CHA*. For ready comparison, the dawn songs of all four males can be heard back-to-back in ♫571. In ♫572 is an example of a common daytime song that is used by many males.

> ## Explore 67: Recognizing individuals by their songs.
>
> The more you listen, the more you will hear, and often without really trying all that hard. To get to know individual birds by their songs, begin with the easier task of focusing on birds who have only one song in their repertoire (see pages 86–87 for a list of candidate species). Listen to song after song from one individual, coming to know the rhythm and overall quality, and especially coming to appreciate the range of variation within his song. Then listen to his neighbor, and you will often find something different that will enable you to tell the two birds apart. The challenge is far greater, of course, when the birds have learned their songs from each other, so that they sing almost identical songs from the same small or large dialect (e.g., some neighboring chipping sparrows, common yellowthroats, indigo buntings, pp. 56–58; or birds who belong to larger dialects, such as white-crowned sparrows, black-capped chickadees, etc., p. 64). The challenge is also great with fly-catchers, as songs vary less from individual to individual than they do in the song-learning songbirds.

EACH SPECIES HAS ITS OWN SONG

The word "species" raises all manner of arguments among professional ornithologists. Just what is a species? How do we decide, for example, if that marsh wren on the West Coast is the same species as the marsh wren who sings so differently on the East Coast (p. 34)? Not easily, not quickly, not without a lot of contemplation! Deciding "species boundaries" is not a hard science, and is influenced by "lumping" and "splitting" philosophies that sway like a pendulum from decade to decade. For songbirds (but not for the suboscine flycatchers), there is the added problem that regional differences in song might not reflect any genetic differences among the birds, but instead reflect only cultural differences (dialects) from place to place. These different dialects in learned song can be no more an indication of different bird species than are different human dialects an indication of many different human species among us.

Here are just a few of the current species dilemmas facing ornithologists. By current, I mean the spring of 2019, but it is entirely possible that some or all of these groups will soon be split, so that we have two ruffed grouse species, two (or more) willow flycatchers, two warbling vireos, and so on.

Ruffed grouse

All male ruffed grouse "drum" in much the same way. They strut up to their special drumming log, rock, or other raised surface, brace themselves with their two feet and their tail, then deliver a wing-flapping routine that is positively thunderous, though the thunder is primarily in frequencies below our range of hearing (see p. 41). Slow initial thumping accelerates into a whir of wings flapping up to 22 times each second, each thump apparently the sound of air rushing into the vacuum beneath the rising wings.

Throughout much of North America, the details of the drum are the same: Two early flaps, then four in a tightly coordinated series, and by five seconds he begins building *gradually* to the climax of over 20 wingbeats per second in second 11, then tapers down as gradually as he rose. He waits a couple of minutes, perhaps listening for distant drummers or approaching females, then drums again. Examples from Massachusetts (♫573) and Idaho (♫574) illustrate the pattern.

A drummer from Vancouver Island in British Columbia is altogether different (♫575). About nine seconds into the drum, he begins what will be *a sudden rush* to the climax, after which he tapers off gradually. Because rooster crows and, no doubt, grouse drums are inborn, these differences in drumming signal a genetically different grouse on Vancouver Island, and possibly elsewhere in the Northwest. Close study may warrant splitting these two different drummers into two different ruffed grouse species.

Explore 68: Discovering new species by their songs.

For many who love birds, discovering a new species is the ultimate quest, the ultimate find. In North America, here's a hot tip: Go to Vancouver Island and record the drumming ruffed grouse there. Even the poorest quality recordings, done with a smart phone, will suffice. If all of the drummers there are different from those of mainland North America, you have a strong case for splitting the "Vancouver Island Ruffed Grouse" into another species. The grouse there and in adjacent areas of British Columbia are already recognized as a distinct subspecies, based on plumage ("dark and brownish with pronounced ventral barring," according to the Birds of North America). Given that the drums of grouse are almost certainly innate and not learned, I place my bets on that grouse being a different species. Better hurry, though, or you'll be scooped!

Willow flycatcher

Because the songs of flycatchers and (most) other suboscines are inborn (see pages 45–46), their songs can be used to detect genetic differences among populations. Such it is with the three songs of the willow flycatcher. Listen to a willow anywhere and you hear its three basic songs (p. 46): the *FITZ-bew* and the *FIZZ-bew,* distinguished by either a sharp *FITZ* or a fizzy *FIZZ* that begins the song, and the simpler, rising *creet.* What fun simply listening to a male chart his progress through these three songs as he delivers them in an unpredictable rhythm and sequence (p. 46, ♫147). Enter the scientist who measures characteristics of the songs, their duration, their frequency, and so much more, and one finds about three identifiable evolutionary groups of these willow flycatchers, one in the East, two in the West. Among the suboscine tropical relatives of our North American flycatchers, these differences in inborn songs are increasingly being used to identify different evolutionary groups, some of which are raised to species status based primarily on differences in inborn song.

In a Missouri (♫576) and an Oregon (♫577) example, can you hear differences that ornithologists might document with close study? Try these back-to-back comparisons in ♫578: (1) the three songs from Missouri, then the three from Oregon; (2) at half speed, the two *creet*s, the two *FITZ-bew*s, and the two *FIZZ-bew*ss, Missouri first, Oregon second; (3) that second sequence at quarter speed. To me, the song elements from Oregon sound lower pitched. The thorough scientist, of course, would make sure that any such differences are regional differences and not just differences between two individuals who happen to be from distant locations.

More about willow flycatchers: "Inborn Songs" (p. 44) and "Song Changes over Evolutionary Time, from Species to Species" (p. 147).

Warbling vireo

Throughout the East, the song of the warbling vireo is unmistakable (also p. 50). It's an undulating roller coaster of a song, building to distinct highs and sinking to distinct lows at a pace in which we can readily chart the progress (*IGGelley, PIGelly, WIGelly, Pig,* said slowly and drawn out; ♫579). But cross the Great Plains and enter

Colorado and an altogether different singer is encountered, one so different that it seems hardly possible it could be a warbling vireo (♫580). Songs are now choppy, leaping from lows to highs, with the song rising and falling so abruptly from one phrase to the next that it doesn't give the same gentle, undulating feeling as does the eastern song. Eastern and western birds differ in other ways, too, such as bill size, body size, tail length, and plumage, suggesting that eventually ornithologists will come to recognize two species here, the eastern warbling vireo and the western warbling vireo.

More about warbling vireos: "Learned Songs of Songbirds, and Babbling" (p. 49).

Bewick's wren

Listen to the Bewick's wren throughout the western United States (they are basically extinct now east of the Mississippi) and you encounter a puzzling variety of songs. Those in southern Arizona are the simplest, consisting mostly of two phrases (♫581); head north to Colorado (♫582) and you find the most complex songs, with up to eight or even ten phrases each. Then consult a range map, and see that birds

along the Pacific coast are isolated from the interior populations. It is no surprise that those coastal songs are different as well; they are intermediate in complexity, and relatively homogeneous up and down the coast (♫583 for an Oregon example, ♫584 from California). Bewick's wrens often use what are presumed to be inborn calls between songs, and those calls between songs in California (♫584) are noticeably different from those in Texas (♫585). It's entirely likely that these differences in songs and calls may signal two or more species among the current Bewick's wren.

For more about the Bewick's wren and the sheer joy of listening to an individual sing through the entire dawn chorus, see "Energized Dawn Singing" (p. 112).

Sedge wren

Everyone agrees that one species of sedge wren occurs throughout North America (♫586). Because the birds are somewhat nomadic, they mix and breed freely with other sedge wrens wherever they go. As a result, they must be genetically uniform across their North American range. Their singing behavior is also homogeneous throughout; birds do not imitate songs, but rather each male improvises a great variety in building a repertoire of a hundred or more songs (pages 48).

If you study these birds as you head south, however, following their range all the way to Tierra del Fuego, you might well conclude, as have some ornithologists, that the currently recognized sedge wren consists of *at least eight different species.* Nowhere was that conclusion more obvious to me than when listening in central Brazil some years ago, where I recorded two sedge wrens matching each other with learned songs (♫587), much as do western marsh wrens (p. 61). Such a major difference in how songs are acquired, whether improvised by North American birds or imitated by birds to the south, suggests that birds from North America and Brazil are entirely different species. Male sedge wrens also learn their songs in Costa Rica (♫588) and the Falkland Islands (♫589), as evidenced by local song dialects there. Different genes guiding different forms of song development demands that we recognize *at least* two species, the improvisers in North America and the imitators to the south.

More about sedge wrens: "Improvised Songs" (p. 47), "Small to Large Repertoires" (p. 83), and "The Music in Birdsong—Improvisation" (p. 159).

Fox sparrow

The fox sparrow illustrates well the challenges that ornithologists face when deciding how to classify birds into species. From Alaska across Canada to Newfoundland one finds the foxy red subspecies; they are fine singers, with each male's *one song* consisting of about a dozen musical notes and rich sliding whistles over two and a half seconds (p. 66; ♫590). Travel from place to place and you readily hear song dialects.

Drop south from central Alaska and you begin to hear very different fox sparrows. Along the coast are the "sooty" brown birds (two examples: ♫591, ♫592), more interior are the "slate-colored" birds (in Montana, ♫593; Idaho, ♫594), and in the Sierras from central Oregon to mid-California, the "large-billed" birds (♫595, ♫596, ♫597). The males in these three groups typically have a repertoire of several different rich and brilliant songs with rapidly slurred whistles and buzzy trills.

It is a fine challenge to try to pick out a unique song and listen for it to recur in a male's performance. In ♫591, for example, try to hear that he sings only two song types, delivered in this sequence: A B A B A B A A. Try ♫592: A A B B A B B (the same letters for different birds do not signify that they are the same songs). The bird in ♫595 begins with A B C D E; ♫596 begins with A B A C A; ♫597 begins A A A A . . . Does that last male really have only one song type? Listen to all 13 songs. How strange!

One species, or two, or four? Based on songs and singing behaviors, I vote for two species, the "red" fox sparrow as one species, the other three lumped into one. Others might vote for four species. The "professional answer" will eventually be handed down by an official committee of the American Ornithological Society, but I doubt the committee's vote will be unanimous.

The songs of the fox sparrow are easily confused with those of the green-tailed towhee. Try listening to ♫598, in which the singer begins A B C D E . . . Go ahead and explore the songs of these two species to try to understand them better.

More about fox sparrows: "Song (and Call) Dialects" (p. 63).

Explore 69: Just what is a "species"?

Ornithologists face one dilemma after another when they try to answer the question "Do these birds represent one or two (or more) species?" We come to take field guides as gospel, but lurking behind so many of the species illustrated there are these kinds of species conundrums. Here is a short list of some field guide species for which ornithologists wonder whether they constitute two or more different species. As you travel, listen up! How do the sounds of these species vary geographically?

Red-shouldered hawk	Swainson's thrush
Red-tailed hawk	Hermit thrush
Northern flicker	Purple finch
Olive-sided flycatcher	Evening grosbeak
Bell's vireo	Savannah sparrow
Common raven	Fox sparrow
Purple martin	White-crowned sparrow
White-breasted nuthatch	Dark-eyed junco
Brown creeper	Orchard oriole
Marsh wren	Common yellowthroat
Bewick's wren	Yellow warbler

SONG CHANGES OVER EVOLUTIONARY TIME, FROM SPECIES TO SPECIES

How fascinating to step out into our world of birdsong and hear it as an evolutionary biologist does, always in the back of our minds asking "Just how did that come to be?" We can think of answers to that question over short time frames, and ask whether a song is inborn, improvised, or learned. But the longer time frame forces us back through evolutionary time, to thinking about "sister species," two or more species who far back in time had a common ancestor. Once upon a time, for example, there was no doubt just one species of meadowlark in North America. Perhaps Pleistocene glaciations isolated one group in the western and another in the eastern half of the continent, and over time the two groups diverged, both genetically and behaviorally, the differences becoming extensive enough that we now classify them as two separate species.

As we listen to closely related pairs or groups of species, we inevitably compare them. We contemplate the similarities in their voices, realizing that the similarities most likely reflect the traits that the birds have retained from their ancestor. We also contemplate the differences, knowing that they have arisen over evolutionary time while the groups were separated and becoming the different species that we recognize today. It's a relatively easy task in our minds to compare pairs of species, but deep down we want to see bigger pictures, and professional ornithologists want answers to questions such as "How did all of those closely related warblers in the genus *Setophaga* come to sing the way they do?" or "How about all of those species in the infamous flycatcher genus *Empidonax?*"

Here we explore five species groups in North America: the tanagers, wood-pewees, kingbirds, meadowlarks, and *Empidonax* flycatchers.

Western tanager, scarlet tanager, summer tanager

The western tanager of the West and the scarlet tanager of the East have nonoverlapping ranges and are remarkably similar in their songs and singing behavior, suggesting that their singing genes have changed relatively little since they shared a common ancestor. During the day, both are the "hoarse robin," or

the robin with the sore throat, offering four to seven burry phrases in each song before pausing several seconds until the next song (western tanager, ♫599; scarlet tanager, ♫600).

During the dawn chorus, they sing continuously but halve the pace, each burry phrase now seeming to stand alone, each phrase now a deliberate enunciation followed by a significant pause, about only one phrase each second instead of two. After every five to ten burry phrases, which among songbirds are expected to be learned in some fashion, the scarlet tanager interjects its innate call, *chip-burr* (♫601), the western tanager its innate call, *pit-er-ick* (♫602), the genetically encoded, unique calls being the best clue from their vocalizations that they are indeed two different species.

Compare now the summer tanager, who had the same ancestor as did others in this *Piranga* genus. By day (♫603), he sings a faster-paced, discrete song without the burriness of the western and scarlet; dawn singing is also slowed (♫604), just like the other tanagers, but the summer tanager does not include his call note. He has a *pit-i-tuck* call note, much like the stuttered call of the other two tanagers, but that's reserved for occasions when the bird appears agitated (♫605).

These comparisons suggest that the scarlet and western tanagers are more closely related to each other than either is to the summer tanager. Taxonomic studies, however, suggest that two other tanagers, the hepatic and flame-colored tanagers, are more closely related to the western tanager than is the scarlet tanager. What fun it would be to explore the singing and calling behaviors of all nine species in the genus *Piranga,* to see how those behaviors map onto the tanager genealogies.

Eastern wood-pewee
Western wood-pewee

These two flycatchers are likely "sister species," meaning they are each other's closest relative, and, like the western and scarlet tanagers, are also separated by the Great Plains into western and eastern species. As with other flycatchers, their songs are believed to be innate, and their songs tell of significant differences in their song genes. By day, the eastern wood-pewee offers a leisurely performance of two different songs: a string of rising, questioning *pee-ah-wee* songs followed by a single, falling answer, *wee-oo*, about five songs each minute (two examples: ♫606, ♫607). During the dawn chorus, he adds a third song, a rising *ah-di-dee,* and also sings at a blistering pace, up to 35 songs per minute (two examples: ♫608, ♫609).

In contrast, the western wood-pewee has but one daytime song, a burry, falling *bzeeyeer* (♫610); for his dawn chorus, he also adds a rising song, his *tswee-tee-teet,* and rapidly alternates his two songs (♫611). Similarities: one song reserved for rapid dawn singing. Differences: eastern species has two daytime songs, the western just one. Are there other similarities and differences?

As with the tanagers, it would be enlightening to compare the singing of these two species in the genus *Contopus* to the nine others in the same genus. There is the olive-sided flycatcher (p. 84) with just one song, but how about the other eight? Just how do they sing? What do they do during the dawn chorus? Throughout the day? How has singing evolved within these closely related species presumably all derived from the same ancestral species?

Eastern kingbird

Western kingbird

Couch's kingbird

Among the dozen or so species of "kingbirds" in the genus *Tyrannus,* these three are not the most closely related, but their singing behaviors, especially at dawn, still reveal singing genes that have been inherited from a common ancestor.

Eastern kingbirds sputter and stutter sharply, *t't'tzeer, t't'tzeer, t'tzeet-zeetzee, t't'tzeer, t't'tzeer, t't'tzeer, t'tzeetzeetzee,* over and over again, as if miserably and repeatedly failing to pronounce the word "explicit," thought one naturalist (♫612). The western kingbird stutters a series of *kip kip kip ki-PIP,* consisting of several low *kip* notes until one is

coupled with a higher and sharper *PIP* note, before he explodes into a *ki-PEEP-PEEP-PEEP-PEEP*, the longer and more pure-toned *PEEP* notes trailing off in both loudness and frequency (♪613).

Both species have a stuttered phrase (*t't'tzeer* and *kip kip kip ki-PIP*, respectively) that ends in a climax (*t'tzeetzeetzee* and *ki-PEEP-PEEP-PEEP-PEEP*), and both species sing with extraordinary energy at dawn. Curiously, the western kingbird at dawn always sings from dense cover, and you can walk up close for a good listen. In contrast, the eastern kingbird, especially in dense populations, takes to the air, circling and singing high above his territory.

Here's a third kingbird, the Couch's kingbird (♪614, briefly; in ♪615 you could chart all 32 minutes of his dawn chorus, from beginning to end!). Listen for the stuttered phrase and the climax.

Western meadowlark

Eastern meadowlark

These look-alike meadowlarks are sibling species who meet in a narrow, overlapping north-south zone in the Great Plains. Where they meet, they sometimes hybridize, with one parent from each species, and, being songbirds, they can also learn each other's songs, creating more than a bit of confusion for us in trying to identify the species of a songster. Well away from the Great Plains, the genetic identity of the singer is unmistakable, the stunning song of western birds (p. 151; ♪616) contrasting sharply with the sliding whistles of the eastern bird (♪617). Say *spring-o'-the-yeeaar* slowly, imagining pure tones gliding smoothly from one to the next, drawing out the lower *yeeaar*, a bit plaintive if heard at any distance, and you have the eastern's song. Western males repeat one of their ten

or so songs several times before switching to another, just as the eastern males do with their hundred or so different songs. How many different songs does the eastern male sing in ♪617? The western male in ♪616?

The only way to reliably identify these two species in the zone of overlap is by their inborn call notes. The eastern meadowlark calls a *dzert*, or longer *dzert-ert-ert-ert-ert-ert* (♪618), the western a *chup, chup,* followed by a dry, rolling chatter, *vicicicicicicic* (♪619).

More about western meadowlarks: "Singing and Calling in Flight" (p. 130) and Explore 70.

> ## Explore 70: The joy of western meadowlark songs.
>
> (See website for details.)

Empidonax flycatchers

Go to your North American field guide and count them; you will find at least ten of these flycatchers in the genus *Empidonax,* but know that there are more in this genus, at least another five to the south, maybe more. No, they don't all look exactly alike, but close enough that I do not try to distinguish them by eye. But how I love to listen to them sing, wondering how all of them came to sing as they do now. Dawn is the best time to listen, of course, and I have made the effort to get to know each of these ten species in the hour before sunrise.

Here they all are, for your listening pleasure. As you listen, think about their journey from their common ancestor, what traits some or all of them have kept, and how they differ from each other. How many different songs, for example, do you hear from each individual, and how does song quality compare among species? The Pacific-slope and Cordilleran flycatchers are most closely related (i.e., have had a common ancestor very recently in geological time) and have almost identical songs, so similar that ornithologists debate whether these two groups should really be called two different species.

Species	Listen here	Number of songs
Yellow-bellied flycatcher	♫630	1
Acadian flycatcher	♫631	1 (plus notes)
Alder flycatcher	♫632	1
Willow flycatcher	♫633	3
Least flycatcher	♫634	1
Hammond's flycatcher	♫635	3
Gray flycatcher	♫636	2
Dusky flycatcher	♫637	3
Pacific-slope flycatcher	♫638	3
Cordilleran flycatcher	♫639	3

More about these flycatchers:

Alder flycatchers: "Inborn Songs" (p. 44).

Least flycatchers: "Song Complexity" (p. 79).

Willow flycatchers: "Inborn Songs" (p. 44) and "Each Species Has Its Own Song" (p. 140).

Explore 71: Song changes over evolutionary time.

The above examples focus on just a few species, but throughout this book are many other examples of closely related species that could be compared. See "Who's Who?" on p. 170, for example, and look for close relatives in the same genus. Here are a few additional comparisons that I find intriguing:

Whip-poor-wills in the genus *Antrostomus* (p. 171)

Sapsuckers in the genus *Sphyrapicus* (p. 173)

Phoebes in the genus *Sayornis* (p. 174)

Vireos in the genus *Vireo* (p. 174)

Chickadees in the genus *Poecile* (p. 175)

Wrens in the genus *Cistothorus* (p. 175)

Thrushes in the genus *Catharus* (p. 176)

Thrashers in the genus *Toxostoma* (p. 176)

Sparrows in the genera *Spizella* and *Zonotrichia* (p. 177)

Warblers in the genera *Geothlypis* and *Setophaga* (p. 178)

Buntings in the genus *Passerina* (p. 178)

9. Music to Our Ears

THE MUSIC IN BIRDSONG

Those of us who know even a little about human music hear features of birdsong that sound "musical" when it strikes our ears. We hear virtuoso performances, such as by a sage thrasher or Townsend's solitaire. We hear stunningly beautiful, pure, musical tones, such as by a white-throated sparrow. We refer to a group of singers, such as can be heard routinely at dawn (e.g., ♫640), as a "chorus" (though the only coordinated effort by the participants is probably to avoid overlapping the songs of others so as not to mask one's own effort).

In turn, birdsong inspires great music, as in Vivaldi's *Four Seasons,* where the return of spring birds and their songs is celebrated by the solo violinist. Birdsong was transcribed directly into the music of Olivier Messiaen. In Respighi's *Pines of Rome,* the songs of a nightingale accompany the orchestra (though, for convenience, a canary is sometimes substituted). Birdsong transcribed into human music can be traced back to the thirteenth century, and by the early 1400s one can find in European compositions the songs of skylarks, song thrushes, and nightingales.

It is best to remember, however, that birdsong is birdsong, and human music is human music, and not to search too deeply for universal features that tell of our distant evolutionary ancestors going back a few hundred million years. Nevertheless, we do hear in birdsong many of the features that are used in human music, and it's worth noting a few of them here.

Ritardando—Canyon wren

RITARDANDO: a gradual decrease in tempo.

The canyon wren's songs are a "cascade of sweet and liquid notes, like the spray of a waterfall and sunshine," wrote the naturalist Arthur Cleveland Bent. These remarkable songs are among the most appreciated in all of North America, the rippling cascade of pure tones dropping down the scale, successive whistles longer and longer, the tempo slowing, the last three or so notes the lowest and slowest. The entire song echoes off the canyon walls and slowly dissipates, the echo fading, the (human) listener left with a sense of "Wow!" and "Please sing it again." The wren often obliges, again and again, then switches to another song, as he has several renditions of this masterpiece in his repertoire.

In ♫641 are 37 songs from one male, compiled from a two-hour period when the male on occasion sang just below me in a collapsed lava tube in northern California (I emphasize: This is not a continuous performance from this male). For decades I have wondered how many different songs a male canyon wren sings, and this is the recording where I can find out. I begin lettering the songs: A B B B C C C D D . . . Yes! I have continued on through the 37 songs and begin to see only repeats, no new songs. A decades-long quest fulfilled. If you are curious, you can have at it yourself to find out how many different songs this male sings! Hint: It's more than four.

When perturbed by an intruding male, a male canyon wren sings on "speed," one song rapidly following another, every one in ritardando mode (♫642).

Accelerando—Field sparrow

ACCELERANDO: a gradual increase in speed.

The routine daytime songs of the field sparrow are also clear, musical whistles (♫643). He begins slowly, almost plaintively, with down-slurred whistled notes, but then he accelerates into a rapid trill, the overall effect that of a "bouncing ball," *teeew teeew teew teew tewtewtewteteteteetititititititi.* He will sing it again, and again, for he has just this one song that he uses throughout the day (but hear him at dawn and you will think he is perhaps confused, now singing backward, his dawn song sometimes sounding like two ritardando songs back-to-back; page 121). What wonderful variety one hears among field sparrows, as some begin their song high and end low, while others begin low and end high; some accelerate gradually, some abruptly, maybe even sounding three-parted, but always an accelerando.

More about field sparrows: "Energized Dawn Singing" (p. 112).

Pitch-shifting—Black-capped chickadee

PITCH-SHIFTING: retaining the same melody in different keys.

Throughout much of North America, one hears the *hey-sweetie* song of the black-capped chickadee (pages 8, 64), but there's something extra in this song. I know exactly where I was standing, back in the mid-1980s, when I first heard it; I doubted my ears at first, but since then this phenomenon has been studied in some depth.

The male typically sings on one pitch for some time (e.g., ♪644), but then abruptly shifts the pitch of his simple song up or down, all the while retaining the same frequency ratio between the *hey* and *sweetie* notes. When excited, as during the dawn chorus, he shifts frequently from one pitch to another, but during the day he often delivers long strings on one pitch and then stops singing before resuming some time later on a different pitch. So your best bet by far to hear the pitch-shifting chickadee is, of course, during the dawn chorus—before sunrise. (You can do it!)

Try listening to these 20 minutes of dawn singing by this Michigan chickadee (♪645). How many pitch shifts do you hear? To detect the shifts more clearly, I excerpted two renditions on each pitch (♪646)—the 11 pitches (!) that he used during those 20 minutes are now readily heard.

More about the black-capped chickadee: "Birds Sing and Call" (p. 6) and "Song (and Call) Dialects" (p. 63).

Explore 72: The curious case of pitch-shifting black-capped chickadees.

Why pitch-shift? And when? Is this form of singing simply another way for a male songbird to interact with singing neighbors? Other song-birds, such as cardinals (p. 59), learn about a dozen different songs from each other and then often match each other during their counter-singing exchanges. Perhaps chickadees play some of the same kinds of games that cardinals do; it's just that when chickadees match, they are matching the pitch of their one *hey-sweetie* song. What fun listening to a community of black-capped chickadees pitch-shifting during the dawn hour, all the while asking questions.

What pitch is he on now? How many songs does he sing on a pitch before shifting? How does that change over time? And his neighbor? Do neighbors pitch-shift at about the same time, and to a similar or a different pitch? Can I detect any patterns? This exercise is facilitated by the chickadees, because during the dawn chorus, male chickadees often gather to countersing near territory boundaries (as do males of so many songbirds), so that two or more birds can easily be heard simultaneously.

Contrasts—Hermit thrush

CONTRAST: the state of being strikingly different, especially in a way that is very obvious.

Each of a male hermit thrush's ten or so strikingly beautiful songs is extraordinary in itself, but it is their sequential delivery that provides the special effect. Listen carefully to the pitch of that introductory whistle, and compare it to the next, and that one to the next, and so on. Hear how he *leaps* from one pitch to another (♫647).

Over evolutionary time, hermit thrushes have had choices. Like song sparrows, they could have come to repeat one of their songs many times before switching to another; listening to hermit thrush songs artificially arranged in this fashion clearly destroys the magnificence of the hermit's performance (♫648). Or the thrushes could have chosen to deliver their songs in a sequence from low to high or high to low frequencies, gradually working up or down the scale, and then starting over—that would have been a rather beautiful effect. Or they could have chosen the next song at random. No, none of that was sufficient. Instead, hermit thrushes have come to sing so that successive songs are not just different from each other, but *especially* different. The effect of heightened contrast in successive songs is pure hermit thrush.

More about the hermit thrush: "Small to Large Repertoires" (p. 83).

Crescendo — Ovenbird

CRESCENDO: a gradual increase in loudness of a song.

That first *teacher* phrase of the ovenbird's song is so quiet that no one could possibly hear it, and by the time the singer ramps up each phrase to the end, the effect is ear-shattering (♫649). Out of curiosity, I measured how the power (in decibels, the way sound engineers measure it) of a *teacher* phrase

changes through one song: the last *teacher* was 45 decibels more powerful than the first. Every 3 decibels is a doubling of power, so that's 15 doublings of power from beginning to end. That can be calculated as 2^{15}, which calculates to 32,768. The last *teacher* phrase was over 30,000 times more powerful than the first! That's impressive. How loud the change seems to our hearing is another matter, and more subjective. About every 10 decibels is perceived as a doubling in loudness, which is still an impressive gain in volume from beginning to end.

More about ovenbirds: "Night Singing" (p. 123) and "Each Individual Has Its Own Song" (p. 137).

Diminuendo — Indigo bunting

DIMINUENDO, OR DECRESCENDO: a gradual decrease in loudness of a song.

Songs of some species trail off toward the end of the song, with the last few phrases noticeably quieter than earlier. The songs of this indigo bunting (♫650) reach a peak in loudness about halfway through the song, and then gradually taper down.

More about indigo buntings: "Learned Songs of Songbirds, and Babbling" (p. 49) and "Big Decisions: When, Where, and from Whom to Learn" (p. 56).

Theme and variations—Brown thrasher

THEME AND VARIATIONS: a recognizable theme that is consistent among all variations on the theme.

The fundamental theme of the brown thrasher lies in his doublets, with more than a thousand variations on the theme of "twoness" (pages 92–93; ♫651); take all brown thrashers rangewide and there must be billions of variations. Edit a northern mockingbird's song by deleting repetitions of a sound until only doublets remain and the song becomes unmistakably brown thrasher. Double each sound in a gray catbird's song performance and, again, they sound like brown thrashers. No other species uses this basic theme of twoness. On average, I should point out, inspection of thrasher sonagrams reveals that many doublets aren't complete, and instead consist of only about one and a half renditions of the component notes.

Every species has its own theme. A wood thrush song has a prelude and a flourish (p. 102); the hermit thrush song, an introductory pure tone followed by a fluty flourish; the white-throated sparrow, long whistled notes followed by shorter ones, and so on. The theme allows us to recognize the species; the variations, all of the nuances by which individuals express themselves within that theme.

More about the brown thrasher: "Why Sing?" (p. 26), "Song (and Call) Matching" (p. 58), and "Small to Large Repertoires" (p. 83).

> ## Explore 73: Themes and variations in birdsong.
>
> Instinctively, we use the "theme and variations" approach as we learn to use songs to identify species. All songs of a given species fall into a recognizable pattern, the theme. Name your species, and then describe how you distinguish the songs of that species from the songs of others. For some species, such as an American redstart, it might be difficult to verbalize what we hear, but we nevertheless do hear a constant "theme," some consistency in a pattern or tempo or quality that identifies the singing bird as a redstart. The variations can be endless, of course, and therein lies the challenge of identifying each singing individual to species!

Improvisation—Sedge wren

IMPROVISE: *to create or arrange a new piece from whatever materials are readily available.*

Improvising musicians extemporaneously create new pieces, often responding to other musicians with techniques or emotions that in the moment feel appropriate, thus inventing new and unique melodies or rhythms or harmonies. It is entirely possible that some birds do exactly that. It wouldn't surprise me if further study showed that a territorial brown thrasher improvises in this manner as he responds to other territorial males during early spring (p. 61).

Mostly, though, when we say that birds "improvise" their songs, we refer to how they acquire their songs: A young bird improvises his own unique songs based on general, pre-existing instructions provided in his DNA, perhaps coupled with some exposure to the songs of accomplished adults. After he has improvised his repertoire of songs during the first year of life, those songs can remain stable throughout adulthood, as with sedge wrens (p. 91; ♫652).

More about sedge wrens: "Improvised Songs" (p. 47), "Small to Large Repertoires" (p. 83), and "Each Species Has Its Own Song" (p. 140).

Dissonance—Varied thrush

DISSONANCE: A disharmonious combination of sounds.

With two voice boxes, birds have choices as to how they use them. When using them simultaneously, harmonies of unimaginable beauty can be created, as with wood thrushes (p. 39). The varied thrush, however, seems to have a different ear for sounds, with the two voices feeling as if they beat against each other, creating dissonance (♫653). During his normal, seemingly slow-motion performance, I grew impatient waiting for his next song, so I tried removing most of his silences between songs (♫654). Nice, but now I slow his songs to half speed (♫655), then to quarter speed (♫656). I now appreciate better not only the dissonance but what seems to be a heightened contrast from one song to the next, as with hermit thrushes (p. 156).

More about varied thrushes: "Not One But Two Voice Boxes" (p. 38).

A metronome—Loggerhead shrike

METRONOME: a device designed to mark exact time by a regularly repeated tick.

Okay, this is a stretch, but it's personal: Listening to a loggerhead shrike (♫657) reminds me of the metronome that my piano teacher would place on top of the piano as I blundered about the keys. Little Donny did not do so well—in spite of how I love bird music, I pretty much failed at the human end of the endeavor. I count (of course I count) 34 songs in the first minute of this shrike's performance, 34 in the second minute, 34 in the third, and *it's the same tick* for the nearly 200 songs over these five and a half minutes. He has other ticks, I know, that he will eventually switch to in his song repertoire, and he no doubt has an adjustable rate of delivery as well, within reason.

> ### Explore 74: Exploring birdsong with human musicians.
>
> Musicians with good ears, especially those with perfect pitch, often hear far more than those of us with normal, untrained ears. You can listen all you want with your own ears, of course, but try borrowing those of a musician friend (along with the friend, of course). Listening with a trained musician can be an ear-opener!

MORE MUSIC TO OUR EARS

Searching analytically for the effects of human music in birdsong is one thing, but appreciating the inescapable, stunning musicality and beauty in it is another. Quite simply, some birds sing in a way that makes any listening human stop and take notice, and simply exclaim "Wow, that's nice!" The songs aren't necessarily extraordinary feats of accomplishment by the birds, but the special songs tend to be low enough and slow enough for our ears to appreciate some inherent magnificence. Here are just a few examples that have universal acclaim, in no particular order.

Purple finch

"Like a sparrow dipped in raspberry juice" is the way Peterson described the looks of the older males, those at least two years old. There is nothing ho-hum about the song either, as over evolutionary time all roughness has been distilled from it, leaving only the highest polish, so rich and mellow, so bright and lively. It's a fast-paced, liquid warble, a rapid series of tumbling musical notes, no two the same, usually rising overall before dropping at the end. And when in flight (♫658), with wings up and floating through the air in full display, the song often continuing for 20 seconds or more, he is truly both a sight and sound to behold. Adding to the overall brilliance of his fine performance is the occasional imitation of the sounds of other species.

Swainson's thrush

One of the most gifted songsters in all of North America, the Swainson's thrush begins humbly, softly, with low throaty notes, but then offers the first of his several spiraling song phrases, each one becoming higher and louder and often longer than the one before, each phrase a microcosm of the larger song that is a smooth, windy, fluty, spiraling progression upward toward the heavens. A few seconds later he delivers another song, and another, offering three to seven subtly different songs in a sequential pattern that's challenging for our human ears to detect. But then slow the songs down (I leave this to you) and one begins to hear the thrush as we think they might hear themselves, this ethereal spirit of spruce and fir woodlands exploiting his two voice boxes to produce a truly virtuoso performance.

In the relative quiet of an Alaskan summer night, listen to the Swainson's thrushes on a stage all their own; one bird sings near, others more distant (♫659). Or try this singer on Chief Joseph Pass, Montana (♫660).

More about the Swainson's thrush: "How Birds Go to Roost and Awake" (p. 108).

Gray-cheeked thrush

Yes, we have listened to them before, singing through the night (p. 128), but once is not enough. Drive Alaska's Denali Highway in season and this is what you enjoy through lingering sunrises, lingering sunsets, and the still of "night" (♫661). The songs seem somewhat ventriloquial, maybe a little more subdued than those of some other thrushes. Perhaps it is the setting that makes the songs so special: the Wrangell Mountains to the southeast, the Alaska Range to the north, Denali itself to the west, wild country extending in all directions. Maybe what so thrills is imagining the sounds at half speed (♫662), with no silence between the songs, listening to the two voices play with each other. At half speed, too, one can more readily hear how many different songs this fellow sings.

Scott's oriole

Nesting among desert palms, arboreal yuccas, and the like, the Scott's oriole is one of the sweetest songsters from the arid Southwest. The first time I heard him I asked myself, "Western meadowlark?" but a closer listen reveals a stunning series of whistled notes that the meadowlark never offers (♫663). Extract the four different songs (♫664) from those five and a half minutes and then listen to them at half speed (♫665). Very nice!

Red-faced warbler

Red-faced warblers are rather striking visually, of course: a bright red face, black bonnet, white nape, slate gray back, white rump patch. But, as with most fine singers, "Seeing is highly overrated," I often say, and you can't really "count them" unless you hear them, and hear them some more.

Listening at length to a red-faced warbler's two- to four-second strings of high, slurred whistles (♫666), one inevitably asks "Does he ever repeat himself?" Clear and sweet, a bit like a yellow warbler who's rambled on far too long, the songs grow on you; try ♪667 for just one of those songs, first at normal, then half speed. And the questions grow, too. Do any other North American warblers sing with this kind of variety and agility? How many different songs occur among the 21 songs of ♪666? How many different songs is he capable of singing? And the curious naturalist by now also comes up with a host of other questions.

Upland sandpiper

This shorebird without a shore has taken to the prairie far from water, bringing to us some of the eloquence of the shorebirds who thrive in the Arctic at the top of our continent. Twice a year many of these shorebirds pass by, to the south in the fall, many of them heading to the far tip of South America, and then northward in spring, but oh so quietly do they pass. Not so the upland sandpiper, who stops to breed in our more southern latitudes.

Whoooooleeeee, wheeelooooo-ooooo! (♪668). The upland sandpiper gurgles pleasantly at first, then rises to the purest of whistles before trailing off and down for the last of three glorious seconds (hear those celebrated details at half speed, ♪669). And in flight, the long neck extending far out front, tail spread, gliding on extended wings with down-turned tips, he offers this enchanting and memorable song before landing with wings up and outstretched over the back before gracefully folding them in place.

Additional examples of exquisite songs and singing permeate this book. Gather up all of your superlatives and then reconsider some of the following, which are among my favorites:

Eastern wood-pewee (p. 148)

Canyon wren (p. 154)

Bewick's wren (p. 167)

Townsend's solitaire (p. 82)

Hermit thrush (p. 88, 156)

Wood thrush (p.39)

Sage thrasher (p. 83)

Cassin's sparrow (p. 96)

Bachman's sparrow (p. 96)

Fox sparrow (p. 66, 145)

White-throated sparrow (p. 67)

Western meadowlark (p. 151)

Rose-breasted grosbeak (p. 32)

Black-headed grosbeak (p. 115)

MUSIC TO MY EARS—AUTHOR'S CHOICE

We all have our favorites, and don't need to justify them. No objectivity is needed. Usually, the bird I am near *now* is my favorite, and they populate this book. But here are a few more that I have saved for last, for no good reason at all.

Common loon

Here is the spirit of the northern wilderness, where fir and spruce point skyward and graceful white birches line rocky shores, and any sound of these birds transports you instantly to the wilds of the North, in splendor and solitude. Most often heard is their tremolo (♫670), sometimes given alone but often in a duet tremolo between mates.

It's a sound like maniacal laughter, used in a variety of circumstances, especially when disturbed, as when boaters approach too closely.

The wail sounds ever so mournful, a few seconds of pure tones that rise or fall, or break suddenly from one frequency to another, much like a wolf howl (♫671). The best guess of ornithologists is that these wails somehow help maintain contact between mates.

In the calm of evening and through the night, listen to a community of loons in early spring shortly after the winter ice has melted on northern lakes, and you will hear the full range of expression (♫672), including the male's yodel song. It is his territorial proclamation, heard under calm conditions well over a mile away; he crouches low in the water, swinging his head from side to side, delivering three or four successive notes that rise higher and higher, at the climax trumpeting out trios of the highest notes.

Bell's vireo

In the predawn darkness, I take stealthy baby steps toward the small bush where the Bell's vireo sings, creeping closer and closer, his songs becoming louder and *louder,* until I stand beside him, exhilarated, singing with him. Each song is an explosive second-and-a-half jumble of rising and falling notes, every three seconds another explosion from this unseen creature hidden deep among the densely packed leaves. Together we ask a question, rising, *cheedle cheedle chee?* followed immediately by an answer, falling, *cheedle cheedle chew!* We complete the question and answer sequence every six seconds, then ask and answer the same again, as if not satisfied. The call and response eventually changes after 50 or so songs, as we sing in packages, favoring the

same question and answer for several minutes until we move on to other questions and answers among our ten-song repertoire. Fueled by his extraordinary energy, I stand with him, listening for the transitions to different packages, marking time by his occasional satisfaction that he has the answer he wants (♫673).

Barn swallow

In the barn swallow's jumbled, bubbling ecstasy of a song (♫674), I hear happiness, perhaps because I am whisked back to a happy childhood spent outdoors, on farms and in barns, the haunt of the oh-so-familiar barn swallow. An objective field guide might describe songs as husky and squeaky, a four-to-five-second rapid twittering

with whines and creaking rattles, but I hear so much more, for as an adult I have now studied the sonagrams and slowed the songs to reveal the details (♫675). He twitters up to six luscious notes per second, each different from the others, bouncing high and low, all so rapidly delivered and varied it sounds as if the notes are sparring with each other. The rising whine now lasts forever, and the creaking rattle on song's end is now a good second of sharp, explosive notes rising up the scale. Wherever I travel, I stop to listen to a singing barn swallow, such as standing on a street corner in Bandera, Texas, with cars driving on by (♫676). When finished singing, he is on the wing, darting out over the fields, catching flying insects, calling *kivik kivik.*

Cliff swallow

Creaking, guttural, and harsh—that's the consensus on these cliff swallows' songs. Ah, but cozy up to a bird singing on the nest to hear what he is really up to (♫677), and then try to imagine an entire colony with hundreds of these singing swallows. I listen to his songs at half speed (♫678) and then enjoy them at quarter speed (♫679), marveling at the extraordinary variety of delightful whines, creaks, and rattles. I smile.

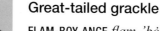

Great-tailed grackle

FLAM-BOY-ANCE *flam-'bȯi-ən(t)s n* : See Great-tailed grackle.

So says my dictionary. They are great to watch, given their flashy personalities, but no encounter with them is complete without a good listen. Alexander Skutch, a tropical naturalist all his life, described the variety of their sounds well: "At one extreme was a rapid sequence of pleasant little tinkling notes; at the other, calls so loud they were best heard at a distance . . . a prolonged slur, between a squawk and a whistle, rising slowly through the tonal scale . . . a rolling or yodeling call . . . the lazy screeching note that reminded me of a gate swinging slowly on rusty hinges . . . a low, undulatory sound like whistling through one's teeth . . . a buglelike call . . . martial and stirring." And all these sounds vary in dialects from place to place. You can be the judge in this series of calls extracted from an hour-long listen (♫680); the effects at half-speed are entrancing (♫681).

More about great-tailed grackles: "How Birds Go to Roost and Awake" (p. 108).

Winter wren

The winter wren has been a favorite since I first heard them in Reese's Swamp in northern Michigan during my ornithology courses in 1968. Twelve years later (1980) I had published a scientific paper on them, showing that the eastern wren was clearly a different species from the western wren, and since then they have been split into the "winter wren" of the East and the "Pacific wren" of the West. The western song is far more complex, but it happens too fast, and as a result sounds percussive to our ears. The eastern song is closer to what my ears can handle, a sweet tumble of pure tones, rising and falling in an undulating pattern all at just the right pace (♫682). Each male typically has two different songs; with each encounter I pause and listen, feeling the undulating rhythm of his song, how he repeats it over and over, until he offers a song with a very different rhythm, his second song.

In ♫683, listen to the two songs of the male in ♫682, first at normal, then half speed, then quarter speed, and then return to ♫682 and try to identify his two songs in real time. For an additional listen, transport yourself to the flanks of Mt. Rogers in Virginia and try to pick out this male's two songs (♫684). When excited during the dawn chorus, a male sings fewer renditions of each song before switching to the other, or even alternates his two songs.

Bewick's wren

In 1969, a half century ago, the Bewick's wren began teaching me so much about birdsong, and about myself. I had been a chemist in college, but found that what lay outside the window of the lab was far more interesting than what was in the test tubes inside. Through 1972 in graduate school, I studied Bewick's wrens, marking hundreds of baby birds with unique combinations of colored bands on their

legs, then following them as they dispersed from home and settled on a territory a mile or more from mom and dad. To discover when and where a young male learned his songs, I recorded the roughly 15 different songs (that was the average; the range was about 13 to 20) from almost every singing male on the William L. Finley National Wildlife Refuge just south of Corvallis, Oregon. Long hours in the field and long hours studying the songs planted the questions for nearly everything I have done since.

And so it came to be that, when a Bewick's wren in Joshua Tree National Park informed us that he was going to sing throughout the dawn chorus from a certain small tree, we came to listen. On successive mornings, for a good half hour each morning, he offered two spectacular performances (♪685, ♪686). It may be true that there are species with greater this or finer that, but objectivity has no place here—on those two mornings, there was no finer place on earth to stand and listen.

More about Bewick's wrens: "Each Species Has Its Own Song" (p. 140).

Explore 76: Get to know this individual wren at Joshua Tree!

(See website for details.)

Explore 77: Develop your own list of favored musicians.

In your own growing list of favored musicians, maybe each is associated with some special time and place in your life, and that's all it takes. I have to confess, though you probably knew all along: These last few "author's choice" birds are special to me, but hardly more so than all the other birds in this book. Listening to me enthuse about birdsong on a ten-week bicycle ride from Virginia to Oregon, my son finally asked, "But how can all birds be the most spectacular?" "It's just the way it is," I explained fully. After 50 years of intensive listening, I find each species and *each individual* fascinating, and the honest answer to the question "Which is your favorite?" is "The bird I'm listening to now."

10. Additional Information

EXTRA! EXTRA! WEB BONUS! MORE BIRDS, MORE SOUNDS!

Having put together a perfectly complete book, covering all the main principles of birdsong, and with abundant examples from birds throughout North America, I am torn, because *there is so much more!* Just think of it: Hundreds more bird species in North America, 10,000 worldwide! So here's something a little extra, for good measure, in no particular order—just a few more listening adventures that intrigued me—to illustrate the Juliet Principle (see dedication page), that "if I listen, there is song everywhere." Details at BirdsongForTheCurious.com!

Eurasian blackbird (♫687)
Red-breasted nuthatch (♫687, ♫689, ♫690)
Common grackle (♫691, ♫692, ♫693)
Black-crested titmouse (♫694)
Blue grosbeak (♫695)
Golden-crowned kinglet (♫696)
Golden-crowned sparrow (♫697)
Seaside sparrow (♫698, ♫699, ♫700, ♫701)
Olive sparrow (♫702)
Rufous-crowned sparrow (♫703, ♫704))
Blue-headed vireo (♫705)
Black-capped vireo (♫706, ♫707, ♫708, ♫709)
Plain chachalaca (♫710)
Montezuma oropendola (♫711, ♫712)

On a whim and a smile, let's go to Australia, land of bowerbirds, kookaburras, honeyeaters, whipbirds, whistlers, fantails, fairy-wrens, and so many more complete strangers to us in North America. Most of the families for these 13 birds are entirely new to this book. All these sounds are, to my mind, vintage Australia! Again, details at Birdsong ForTheCurious.com.

Plumed whistling duck (♫713)
Tooth-billed bowerbird (♫714, ♫715, ♫716, ♫717, ♫718)
Green catbird (♫719)
Red-backed fairy-wren (♫720)
Bell miner (♫721)
Scarlet honeyeater (♫722)
Chowchilla (♫723, ♫724)
Eastern whipbird (♫725, ♫726)
Golden whistler (♫727, ♫728, ♫729)
Rufous whistler (♫730)
Rufous fantail (♫731)
Willie wagtail (♫732)
Forest raven (♫733)
Tasmanian devil (♫734; OK, not a bird, but what a fine sound from Down Under!)

WHO'S WHO?

As with any gathering, people or birds, we want to know who's related to whom, and just how they all fit in. Orders, families, common names, and scientific names are all listed here, with the more ancient groups listed first (e.g., ducks and geese), the groups that have evolved more recently listed last (cardinals, grosbeaks, etc.). Birds in the same genus are more closely related to each other than to other species listed in the same family, and those in the same family are more closely related to each other than to those in other families, and so on. In other words, those in the same genus shared a common ancestor more recently than they did with others in the same family, and when we are interested in understanding how songs change over evolutionary time, we want to know these kinds of relationships. Names used here are from the Checklist of North & Middle American Birds published online by the

American Ornithological Society (see http://checklist.aou.org/). (Birds listed in parentheses are mentioned in the book only in passing and have no dedicated recording of them.

Anseriformes
　　Anatidae
　　　　Canada goose *Branta canadensis*
　　　　(Wood duck *Aix sponsa*)
　　　　American wigeon *Mareca americana*
　　　　Mallard *Anas platyrhynchos*
　　　　Ruddy duck *Oxyura jamaicensis*
Galliformes
　　Cracidae
　　　　Plain chachalaca *Ortalis vetula*
　　Odontophoridae
　　　　Northern bobwhite *Colinus virginianus*
　　　　California quail *Callipepla californica*
　　　　Red junglefowl *Gallus gallus*
　　Phasianidae
　　　　Ruffed grouse *Bonasa umbellus*
　　　　Wild turkey *Meleagris gallopavo*
Podicipediformes
　　Podicipedidae
　　　　Pied-billed grebe *Podilymbus podiceps*
Columbiformes
　　Columbidae
　　　　Rock pigeon *Columba livia*
　　　　Eurasian collared-dove *Streptopelia decaocto*
　　　　Mourning dove *Zenaida macroura*
Caprimulgiformes
　　Caprimulgidae
　　　　Common nighthawk *Chordeiles minor*
　　　　(Common poorwill *Phalaenoptilus nuttallii*)
　　　　Chuck-will's-widow *Antrostomus carolinensis*
　　　　Eastern whip-poor-will *Antrostomus vociferus*
　　　　Mexican whip-poor-will *Antrostomus arizonae*
Apodiformes
　　Apodidae
　　　　Chimney swift *Chaetura pelagica*

Trochilidae
(Ruby-throated hummingbird *Archilochus colubris*)
Anna's hummingbird *Calypte anna*
Broad-tailed hummingbird *Selasphorus platycercus*
Gruiformes
Rallidae
Virginia rail *Rallus limicola*
Sora *Porzana carolina*
Aramidae
Limpkin *Aramus guarauna*
Gruidae
Sandhill crane *Antigone canadensis*
Charadriiformes
Charadriidae
Killdeer *Charadrius vociferus*
Scolopacidae
Upland sandpiper *Bartramia longicauda*
Long-billed curlew *Numenius americanus*
American woodcock *Scolopax minor*
Wilson's snipe *Gallinago delicata*
Willet *Tringa semipalmata*
Alcidae
(Pigeon guillemot *Cepphus columba*)
Laridae
Laughing gull *Leucophaeus atricilla*
Western gull *Larus occidentalis*
Gaviiformes
Gaviidae
Common loon *Gavia immer*
Procellariiformes
Procellariidae
Sooty shearwater *Ardenna grisea*
Suliformes
Phalacrocoracidae
Double-crested cormorant *Phalacrocorax auritus*
Anhingidae
Anhinga *Anhinga anhinga*
Pelecaniformes
Ardeidae

    (American bittern *Botaurus lentiginosus*)
    Great blue heron *Ardea herodias*
    (Black-crowned night-heron *Nycticorax nycticorax*)
Accipitriformes
    Accipitridae
        Red-shouldered hawk *Buteo lineatus*
        Broad-winged hawk *Buteo platypterus*
        Red-tailed hawk *Buteo jamaicensis*
Strigiformes
    Strigidae
        (Western screech-owl *Megascops kennicottii*)
        (Eastern screech-owl *Megascops asio*)
        Great horned owl *Bubo virginianus*
        Northern hawk owl *Surnia ulula*
        Barred owl *Strix varia*
Coraciiformes
    Alcedinidae
        (Belted kingfisher *Megaceryle alcyon*)
Piciformes
    Picidae
        Red-headed woodpecker *Melanerpes erythrocephalus*
        Red-bellied woodpecker *Melanerpes carolinus*
        Yellow-bellied sapsucker *Sphyrapicus varius*
        Red-naped sapsucker *Sphyrapicus nuchalis*
        Red-breasted sapsucker *Sphyrapicus ruber*
        (Downy woodpecker *Dryobates pubescens*)
        (Hairy woodpecker *Dryobates villosus*)
        Northern flicker *Colaptes auratus*
        Pileated woodpecker *Dryocopus pileatus*
Falconiformes
    Falconidae
        American kestrel *Falco sparverius*
Passeriformes
    Tyrannidae
        Great crested flycatcher *Myiarchus crinitus*
        Couch's kingbird *Tyrannus couchii*
        Western kingbird *Tyrannus verticalis*
        Eastern kingbird *Tyrannus tyrannus*
        Olive-sided flycatcher *Contopus cooperi*

Western wood-pewee *Contopus sordidulus*
Eastern wood-pewee *Contopus virens*
Yellow-bellied flycatcher *Empidonax flaviventris*
Acadian flycatcher *Empidonax virescens*
Alder flycatcher *Empidonax alnorum*
Willow flycatcher *Empidonax traillii*
Least flycatcher *Empidonax minimus*
Hammond's flycatcher *Empidonax hammondii*
Gray flycatcher *Empidonax wrightii*
Dusky flycatcher *Empidonax oberholseri*
Pacific-slope flycatcher *Empidonax difficilis*
Cordilleran flycatcher *Empidonax occidentalis*
Black phoebe *Sayornis nigricans*
Eastern phoebe *Sayornis phoebe*
Say's phoebe *Sayornis saya*
Laniidae
Loggerhead shrike *Lanius ludovicianus*
Vireonidae
Black-capped vireo *Vireo atricapilla*
White-eyed vireo *Vireo griseus*
Bell's vireo *Vireo bellii*
(Hutton's vireo *Vireo huttoni*)
Yellow-throated vireo *Vireo flavifrons*
(Cassin's vireo *Vireo cassinii*)
Blue-headed vireo *Vireo solitarius*
(Plumbeous vireo *Vireo plumbeus*)
Philadelphia vireo *Vireo philadelphicus*
Warbling vireo *Vireo gilvus*
Red-eyed vireo *Vireo olivaceus*
Corvidae
Pinyon jay *Gymnorhinus cyanocephalus*
Steller's jay *Cyanocitta stelleri*
Blue jay *Cyanocitta cristata*
California scrub-jay *Aphelocoma californica*
(Clark's nutcracker *Nucifraga columbiana*)
Black-billed magpie *Pica hudsonica*
American crow *Corvus brachyrhynchos*
Common raven *Corvus corax*

Alaudidae
    Horned lark *Eremophila alpestris*
Hirundinidae
    Purple martin *Progne subis*
    Tree swallow *Tachycineta bicolor*
    Cliff swallow *Petrochelidon pyrrhonota*
    Barn swallow *Hirundo rustica*
Paridae
    Carolina chickadee *Poecile carolinensis*
    Black-capped chickadee *Poecile atricapillus*
    (Mountain chickadee *Poecile gambeli*)
    Chestnut-backed chickadee *Poecile rufescens*
    (Boreal chickadee *Poecile hudsonicus*)
    Oak titmouse *Baeolophus inornatus*
    (Juniper titmouse *Baeolophus ridgwayi*)
    Tufted titmouse *Baeolophus bicolor*
    Black-crested titmouse *Baeolophus atricristatus*
Aegithalidae
    Bushtit *Psaltriparus minimus*
Sittidae
    Red-breasted nuthatch *Sitta canadensis*
    White-breasted nuthatch *Sitta carolinensis*
Certhiidae
    (Brown creeper *Certhia Americana*)
Troglodytidae
    Rock wren *Salpinctes obsoletus*
    Canyon wren *Catherpes mexicanus*
    House wren *Troglodytes aedon*
    Pacific wren *Troglodytes pacificus*
    Winter wren *Troglodytes hiemalis*
    Sedge wren *Cistothorus platensis*
    Marsh wren *Cistothorus palustris*
    Carolina wren *Thryothorus ludovicianus*
    Bewick's wren *Thryomanes bewickii*
    Cactus wren *Campylorhynchus brunneicapillus*
Regulidae
    Golden-crowned kinglet *Regulus satrapa*
    Ruby-crowned kinglet *Regulus calendula*

Sylviidae
   Wrentit *Chamaea fasciata*
Turdidae
   Eastern bluebird *Sialia sialis*
   Mountain bluebird *Sialia currucoides*
   Townsend's solitaire *Myadestes townsendi*
   Veery *Catharus fuscescens*
   Gray-cheeked thrush *Catharus minimus*
   Swainson's thrush *Catharus ustulatus*
   Hermit thrush *Catharus guttatus*
   Wood thrush *Hylocichla mustelina*
   American robin *Turdus migratorius*
   Eurasian blackbird *Turdus merula*
   Varied thrush *Ixoreus naevius*
Mimidae
   Gray catbird *Dumetella carolinensis*
   Brown thrasher *Toxostoma rufum*
   California thrasher *Toxostoma redivivum*
   Sage thrasher *Oreoscoptes montanus*
   Northern mockingbird *Mimus polyglottos*
Sturnidae
   European starling *Sturnus vulgaris*
Bombycillidae
   Cedar waxwing *Bombycilla cedrorum*
Passeridae
   House sparrow *Passer domesticus*
Fringillidae
   Evening grosbeak *Coccothraustes vespertinus*
   (Pine grosbeak *Pinicola enucleator*)
   House finch *Haemorhous mexicanus*
   Purple finch *Haemorhous purpureus*
   Cassin's finch *Haemorhous cassinii*
   Common redpoll *Acanthis flammea*
   Red crossbill *Loxia curvirostra*
   (Cassia crossbill *Loxia sinesciuris*)
   (Lesser goldfinch *Spinus psaltria*)
   (Lawrence's goldfinch *Spinus lawrencei*)
   American goldfinch *Spinus tristis*

Calcariidae
McCown's longspur *Rhynchophanes mccownii*
Passerellidae
Olive sparrow *Arremonops rufivirgatus*
Green-tailed towhee *Pipilo chlorurus*
Spotted towhee *Pipilo maculatus*
Eastern towhee *Pipilo erythrophthalmus*
Rufous-crowned sparrow *Aimophila ruficeps*
Cassin's sparrow *Peucaea cassinii*
Bachman's sparrow *Peucaea aestivalis*
American tree sparrow *Spizelloides arborea*
Chipping sparrow *Spizella passerina*
Clay-colored sparrow *Spizella pallida*
Brewer's sparrow *Spizella breweri*
Field sparrow *Spizella pusilla*
Vesper sparrow *Pooecetes gramineus*
Lark sparrow *Chondestes grammacus*
Black-throated sparrow *Amphispiza bilineata*
Lark bunting *Calamospiza melanocorys*
Savannah sparrow *Passerculus sandwichensis*
Henslow's sparrow *Centronyx henslowii*
Seaside sparrow *Ammospiza maritima*
Fox sparrow *Passerella iliaca*
Song sparrow *Melospiza melodia*
White-throated sparrow *Zonotrichia albicollis*
White-crowned sparrow *Zonotrichia leucophrys*
Golden-crowned sparrow *Zonotrichia atricapilla*
Dark-eyed junco *Junco hyemalis*
Icteridae
Yellow-breasted chat *Icteria virens*
Yellow-headed blackbird *Xanthocephalus xanthocephalus*
Bobolink *Dolichonyx oryzivorus*
Eastern meadowlark *Sturnella magna*
Western meadowlark *Sturnella neglecta*
Montezuma oropendola *Psarocolius montezuma*
Orchard oriole *Icterus spurius*
Baltimore oriole *Icterus galbula*
Scott's oriole *Icterus parisorum*
Red-winged blackbird *Agelaius phoeniceus*

Brown-headed cowbird *Molothrus ater*
Common grackle *Quiscalus quiscula*
Great-tailed grackle *Quiscalus mexicanus*
Parulidae
Ovenbird *Seiurus aurocapilla*
(Northern waterthrush *Parkesia noveboracensis*)
Tennessee warbler *Oreothlypis peregrina*
Nashville warbler *Oreothlypis ruficapilla*
Connecticut warbler *Oporornis agilis*
(MacGillivray's warbler *Geothlypis tolmiei*)
Common yellowthroat *Geothlypis trichas*
American redstart *Setophaga ruticilla*
Magnolia warbler *Setophaga magnolia*
Yellow warbler *Setophaga petechia*
Chestnut-sided warbler *Setophaga pensylvanica*
Black-throated blue warbler *Setophaga caerulescens*
Red-faced warbler *Cardellina rubrifrons*
Cardinalidae
(Hepatic tanager *Piranga flava*)
Summer tanager *Piranga rubra*
Scarlet tanager *Piranga olivacea*
Western tanager *Piranga ludoviciana*
(Flame-colored tanager *Piranga bidentata*)
Northern cardinal *Cardinalis cardinalis*
Pyrrhuloxia *Cardinalis sinuatus*
Rose-breasted grosbeak *Pheucticus ludovicianus*
Black-headed grosbeak *Pheucticus melanocephalus*
Blue grosbeak *Passerina caerulea*
Lazuli bunting *Passerina amoena*
Indigo bunting *Passerina cyanea*
Painted bunting *Passerina ciris*
Dickcissel *Spiza americana*

How to hear and see birdsong

For seeing birdsong, there is no better (free!) program than Raven Lite, available from the Bioacoustics Research Program at the Cornell Laboratory of Ornithology (available here: http://ravensoundsoftware. com/software/raven-lite/). Downloading Raven Lite to your computer can expand enormously your appreciation for birds and their sounds.

Every one of the sound files that I offer can be downloaded and imported into Raven Lite, allowing you to alert your ears to what you can see with your eyes. Stretch a sound out on the computer monitor and see the details, marveling at how precision breathing through two voice boxes, all controlled by those song control centers in the brain, produces such intricate sounds. Slow the songs down to half, quarter, eighth, or whatever speed you want, pretending you are the ultimate bird with the finest of hearing. Play the sound backward and hear the reverberation that now *precedes* instead of *follows* the sound, and hear the double strike in sapsucker drumming. Oh, and the numbers you can collect! You can measure almost anything you choose—time events to a fraction of a second, or measure frequency accurately.

Here is some additional help on how to visualize bird sounds: (1) In *The Singing Life of Birds* (2005), I provide many sonagrams with the sounds to match them, enabling the eyes and ears to work in concert; (2) the introductory section titled "Visualizing Sound" in Nathan Pieplow's *Peterson Field Guide to Bird Sounds of Eastern North America* (2017) provides abundant information on how to think about the qualities of sound that you hear, and how those qualities might be visualized in sonagrams; (3) the website for the Macaulay Library at Cornell University (https://search.macaulaylibrary.org/) allows you to listen to your favorite bird sounds as sonagrams automatically play across your computer screen.

Recording birdsong

Soon you won't be content using the recordings that others supply for you, and you will want to make your own. With modern technology, that undertaking is increasingly simple, and increasingly inexpensive. On this book's website, I offer ideas on how to get started.

Additional resources

Websites, books, sound archives—a number of resources for the curious birdsong naturalist are provided on this book's website.

Notes

Throughout the book, I occasionally refer to information that feels as if it needs justification, or a reference to give someone credit, beyond the usual information that would be found in the Birds of North America accounts. Here are a few of those references.

3   **One of my favorite book titles** Lawrence Weschler, *Seeing Is Forgetting the Name of the Thing One Sees* (Oakland, CA: University of California Press, 1982).

8   **the more *dee* notes, the more dangerous the predator** C. N. Templeton, E. Greene, and K. Davis, "Allometry of Alarm Calls: Black-capped Chickadees Encode Information about Predator Size," *Science* 308 (June 24, 2005): 1934–1937.

9   **white-breasted nuthatch is believed to have just two songs** G. Ritchison, "Vocalizations of the White-breasted Nuthatch," *Wilson Bulletin* 95 (1983): 440–451.

12   ***our best human attempt* to place bird sounds into two bins** For the difficulty of defining "song," see D. A. Spector, "Definition in Biology: The Case of 'Bird Song,'" *Journal of Theoretical Biology* 168 (4) (1995): 373–381.

20   **she often sings . . . in response to her mate singing nearby** S. L. Halkin, "Nest-Vicinity Song Exchanges May Coordinate Biparental Care in Northern Cardinals," *Animal Behaviour* 54 (July 1997): 189–198.

27   **elevating the southwestern birds to full species status in 2010** R. T. Chesser, R. C. Banks, F. K. Barker, C. Cicero, J. L. Dunn, A. W. Kratter, I. J. Lovette, P. C. Rasmussen Jr., J. V. Remsen, J. D. Rising, D. F. Stotz, and K. Winker, "Fifty-first Supplement to the American Ornithologists' Union *Checklist of North American Birds*," *Auk* 127 (3) (2010): 726–744. Periodically, as illustrated by this 2010 article, the American Ornithologists' Union, now the American Ornithological Society, updates its opinions about the taxonomic status of North American birds.

30 **"wail of despairing agony"** from W. L. Dawson, *The Birds of California: A Complete, Scientific and Popular Account of the 580 Species and Subspecies of Birds found in the State,* vol. 2 (San Diego, CA: South Moulton Company, 1923).

32 **"full of life and animation . . . poured forth with great fervor"** A. C. Bent, *Life Histories of North American Cardinals, Grosbeaks, Buntings, Towhees, Finches, Sparrows, and Allies,* Bulletin of the United States National Museum 237 (Washington, D.C., Smithsonian Institution, 1968).

40 **begins with one side of the syrinx and seamlessly transitions to the other half to complete his slur.** The late Roderick A. Suthers at Indiana University has contributed enormously to our understanding of how birds use their two voice boxes in "ordinary singing."

41 **"one of the poorest vocal efforts of any bird"** R. T. Peterson, *A Field Guide to the Birds: Giving Field Marks of All Species Found East of the Rockies* (Boston: Houghton Mifflin, 1947).

44 **bellbirds in the genus *Procnias* [learn their songs]** D. Kroodsma, D. Hamilton, J. E. Sanchez, B. E. Byers, H. Fandiño-Mariño, D. W. Stemple, J. M. Trainer, and G. V. N. Powell, "Behavioral Evidence for Song Learning in the Suboscine Bellbirds (*Procnias* spp.; Cotingidae). *Wilson Journal of Ornithology* 125 (March 2013): 1–14.

53 *fire fire where where heeere my my run run run faster faster safe safe pheeeewwww* I love David Sibley mnemonics. D. A. Sibley, *The Sibley Guide to Birds* (New York: Alfred A. Knopf, 2000).

72 **promoted to full species status: the Cassia crossbill** C. W. Benkman, J. W. Smith, P. C. Keenan, T. L. Parchman, and L. Santisteban, "A New Species of the Red Crossbill (Fringillidae: *Loxia*) from Idaho," *The Condor* 111 (2009): 169–176.

131 **her *pill-will-willet* is said to be somewhat "less musical and flatter"** Birds of North America, https://birdsna.org/. I once thought that female cardinal songs were inferior to those of males, too, but I was wrong (p. 20). Female willet songs need more study.

138 **Back in 1959, two enterprising ornithologists showed** J. S. Weeden and J. B. Falls, "Differential Responses of Male Ovenbirds to Recorded Songs of Neighboring and More Distant Individuals," *Auk* 76 (July 1, 1959): 343–351.

149 **repeatedly failing to pronounce the word "explicit," thought one naturalist** Walter Faxon in A. C. Bent, *Life Histories of North American Flycatchers, Larks, Swallows, and Their Allies,* Bulletin of the United States National Museum 179 (Washington, DC: Smithsonian Institution, 1942).

154 **". . . like the spray of a waterfall and sunshine," wrote the naturalist Arthur Cleveland Bent.** A. C. Bent, *Life Histories of North American Nuthatches, Wrens, Thrashers, and Their Allies,* Bulletin of the United States National Museum 195 (Washington, DC: Smithsonian Institution, 1942).

159 **heightened contrast from one song to the next** So concluded Carl Whitney: C. L. Whitney, "Patterns of Singing in the Varied Thrush: II. A Model of Control," *Zeitschrift für Tierpsychologie* 57 (1981): 141–162.

161 **"Like a sparrow dipped in raspberry juice"** R. T. Peterson, *Peterson Field Guide to Birds of North America* (Boston: Houghton Mifflin, 2008).

166 **Alexander Skutch, a tropical naturalist all his life** A. F. Skutch, *Orioles, Blackbirds, and Their Kin: A Natural History* (Tucson, AZ: University of Arizona Press, 1996).

Photo Credits

Page numbers are in **bold**, followed by a dash and the number of the image on that page, with images numbered left to right, top to bottom.

Wil Hershberger: **7**–5; **9**–1; **10**–2; **13**–2; **17**–5; **19**–2; **21**–1; **24**–1; **27**–1; **28**–1; **29**–1; **30**–1; **35**–2; **39**–1; **40**–1; **41**–2; **45**–1; **48**–2; **50**–1; **51**–2, 3; **58**–1; **59**–1; **60**–1; **66**–1; **67**–1, 2; **78**–2; **82**–2; **84**–3; **86**–1; **87**–1; **88**–1; **94**–1; **96**–2; **98**–1; **105**–1; **109**–1; **110**–1; **113**–1; **114**–2; **117**–1; **120**–1; **121**–1, 2; **123**–1; **124**–2; **127**–1; **130**–1; **134**–1, 2; **135**–1, 2; **139**–2; **148**–1; **154**–2; **158**–1; **160**–1; **161**–1, 2; **163**–1

Brian E. Small: **7**–1, 3, 6, 8, 9, 10, 11; **8**–1; **10**–1; **13**–1; **15**–3; **17**–3, 4; **18**–1; **20**–1, 2, 3; **22**–3; **35**–1; **36**–2; **39**–2; **42**–1; **45**–3; **46**–1; **48**–1; **49**–1; **53**–3, 4; **60**–3; **62**–1; **79**–1; **80**–1; **83**–1; **84**–1; **88**–2; **92**–1; **95**–1, 2; **98**–2; **102**–2; **107**–1; **110**–2; **126**–2; **128**–1; **143**–1; **144**–1, 2; **149**–1; **150**–1, 2; **151**–1, 2, 3; **155**–1; **159**–1, 2; **162**–2, 3; **165**–1; **167**–2

Robert Royse: **17**–1, 2; **23**–1; **29**–2; **32**–1; **33**–2; **34**–1, 2; **36**–1; **37**–1; **44**–1; **51**–1; **57**–2, 3; **63**–1; **65**–1; **66**–2; **77**–1; **82**–1; **84**–2; **86**–2; **90**–1; **91**–1; **96**–1; **100**–1; **101**–1; **104**–2; **105**–2; **106**–1; **109**–2; **110**–3; **115**–2; **128**–2; **142**–1; **145**–1; **149**–3; **150**–3, 4; **154**–1; **162**–1; **166**–2; **167**–1

Marie Read: **7**–2, 4, 7, 12; **11**–1, 2; **12**–1, 2; **15**–1, 2; **16**–1; **18**–2, 3; **19**–1, 3; **22**–1, 2; **23**–2; **24**–2; **25**–1; **27**–2; **31**–1; **32**–2; **42**–2; **46**–2; **53**–1, 2; **57**–1; **60**–2; **68**–1; **72**–1; **73**–1; **76**–1; **102**–1; **104**–1; **108**–1; **115**–1; **117**–2; **126**–1; **129**–1; **147**–1; **148**–2; **157**–1; **166**–1

John Van de Graaff: **14**–1; **27**–3; **28**–2; **54**–1, 2; **55**–1; **62**–2; **64**–2; **69**–1; **70**–1; **80**–2; **81**–2; **89**–1; **111**–1; **114**–1; **125**–1; **131**–1; **132**–1, 2; **133**–1; **138**–1; **149**–2; **156**–1; **157**–2; **164**–1; **165**–2;

Laure W. Neish: **13**–3; **24**–3; **31**–2; **41**–1; **45**–2; **52**–1; **64**–1; **72**–2; **78**–1; **81**–1; **133**–2; **139**–1; **141**–1; **161**–3

Amanda Morgan Riley: **33**–1; **38**–1

Donald Kroodsma: **43**–1; **137**–1

J. Turner MD/VIREO: **124**–1

Index

Page numbers in **bold** indicate a focus on that species or concept, usually with sounds for listening.

Acanthis flammea. See redpoll, common

Accipitridae, 173

Accipitriformes, 173

Aegithalidae, 175

Agelaius phoeniceus. See blackbird, red-winged

Ailuroedus crassirostris. See catbird, green

Aimophila ruficeps. See sparrow, rufous-crowned

Aix sponsa. See duck, wood

Alaudidae, 174

Alcedinidae, 173

Alcidae, 172

alligator, 5

American Birding Association, 3

American Ornithological Society, 170

Ammospiza maritima. See sparrow, seaside

amphibian, 5

Amphispiza bilineata. See sparrow, black-throated

Anas platyrhynchos. See mallard

Anatidae, 171

anhinga, 5, 172

  female song, duets, **17**

Anhinga anhinga. See anhinga

Anhingidae, 172

Anseriformes, 171

antbird, 6

Antigone canadensis. See crane, sandhill

Antrostomus, 152

arizonae. See whip-poor-will, Mexican

carolinensis. See chuck-will's-widow

vociferus. See whip-poor-will, eastern

Aphelocoma californica. See scrub-jay, California

Apodidae, 171

Aramidae, 172

Aramus guarauna. See limpkin

Archilochus colubris. See hummingbird, ruby-throated

Ardea herodias. See heron, great blue

Ardeidae, 172

Ardenna grisea. See shearwater, sooty

Arremonops rufivirgatus. See sparrow, olive

Audubon Society, 3

Australia, sounds of, 170

awaking and roosting, **108–12**

babbling (subsong), **49–55**

bachelor males and song, **26–29**

Baeolophus

  atricristatus. See titmouse, black-crested

  bicolor. See titmouse, tufted

  inornatus. See titmouse, oak

  ridgwayi. See titmouse, juniper

Bartramia longicauda. See sandpiper, upland

beauty in birdsong, 1

bellbirds (*Procnias* spp.), 44, 50

birds
defined, 5
evolution of, 5
Birds of North America, 4, 21
bittern, 5
American, 173
blackbird, 3
Eurasian, 169, 176
red-winged, 88, 97, 177
call matching, **71**
dawn singing, **117**, **119**
dialects, **70–71**
female song, duets, **18–19**
individuality, **119**
polygyny and song, 29
yellow-headed, 88, 90, 136, 177
Why sing?, **30**
bluebird
eastern, 43, 73–74, 176
dawn singing, **114**
mountain, 113, 176
flight songs, **133**
bobolink, 88, 136, 177
dialects, **68**
polygyny and song, 29
bobwhite, northern, 74, 100–1,
171
inborn song, **44**
Bombycilla cedrorum. See waxwing,
cedar
Bombycillidae, 176
Bonasa umbellus. See grouse, ruffed
Bonus sounds
Australia, 170
North America, 169
Botaurus lentiginosus. See bittern,
American
bowerbird, tooth-billed, 170
brain and song, **32–37**
Branta canadensis. See goose, Canada
Bubo virginianus. See owl, great
horned
bunting
indigo, 50, 87, 152, 178

dialects, 140
diminuendo, **157**
mini-dialects, 57, 63
socially monogamous, 26
song learning decisions,
57
subsong, 50, **53**
lark, 114, 177
flight songs, **134–135**
lazuli, 87, 152, 178
mini-dialects, 57, 63
song learning decisions, **57**
painted, 152, 178
song complexity, **81**
bushtit, 175
no song?, **107**
Buteo
jamaicensis. See hawk, red-tailed
lineatus. See hawk, red-shouldered
platypterus. See hawk,
broad-winged

Calamospiza melanocorys. See
bunting, lark
Calcariidae, 176
call vs. song, **6–15**
Callipepla californica. See quail,
California
Calypte anna. See hummingbird,
Anna's
Campylorhynchus brunneicapillus.
See wren, cactus
Caprimulgidae, 171
Caprimulgiformes, 171
Cardellina rubrifrons. See warbler,
red-faced
cardinal, northern, 74, 93, 97, 113,
170, 178
awaking, **112**
female song, duets, **19**, **20**, 21
song matching, **59**, **60**
song repertoire, **89**
two voices, **40**
Cardinalidae, 178

Cardinalis
   cardinalis. See cardinal, northern
   sinuatus. See pyrrhuloxia
Carduelinae, 72, 133
cat, 8, 78
catbird
   green, 170
   gray, 73, 91, 100, 106, 158,
     176
     improvised song, **49**
     mimicry, 78
     repertoire, 37
     song and call, **11**
Catharus, 152
   fuscescens. See veery
   guttatus. See thrush, hermit
   minimus. See thrush, gray-cheeked
   ustulatus. See thrush, Swainson's
Catherpes mexicanus. See wren,
   canyon
Centronyx henslowii. See sparrow,
   Henslow's
Cepphus columba. See guillemot,
   pigeon
Certhia Americana. See creeper,
   brown
Certhiidae, 175
chachalaca, plain, 169, 171
Chaetura pelagica. See swift, chimney
Chamaea fasciata. See wrentit
Charadriidae, 172
Charadriiformes, 172
Charadrius vociferus. See killdeer
chat, yellow-breasted, 101, 136,
   177
   mimicry, 78
   repertoire delivery, **100–01**
chickadee, 3, 6
   black-capped, 76, 87, 92, 97, 106,
     113, 152, 175
     dialects, **64–65**, 140
     pitch-shifting, **155**
     song and call, **8**
   boreal, 8, 152, 175

Carolina, 8, 97, 106, 113, 152,
   175
   song repertoire, **87**
chestnut-backed, 8, 113, 152, 175
   no song?, **106**
mountain, 8, 106, 152, 175
chick-a-dee calls, 8
chicken, 5, 36, 137. *See also*
   junglefowl, red
Chondestes grammacus. See sparrow,
   lark
Chordeiles minor. See nighthawk,
   common
chowchilla, 170
Christmas Bird Count, 125
chuck-will's-widow, 171
   night singing, **124**
   Why sing?, 28
Cistothorus, 152
   palustris. See wren, marsh
   platensis. See wren, sedge
Coccothraustes vespertinus. See
   grosbeak, evening
Colaptes auratus. See flicker, northern
Colinus virginianus. See bobwhite,
   northern
collared-dove, Eurasian, 171
   inborn song, **45**
Columba livia. See pigeon, rock
Columbidae, 171
Columbiformes, 171
complexity of songs, **79–83**
Contopus, 84, 149
   cooperi. See flycatcher, olive-sided
   sordidulus. See wood-pewee,
     western
   virens. See wood-pewee, eastern
Coraciiformes, 173
cormorant, double-crested, 5, 172
   song vs. call, **13**
Corvidae, 59, 105, 174
Corvus
   brachyrhynchos. See crow,
     American

corax. See raven, common
tasmanicus. See raven, forest
courtship songs, **30–32**
cowbird, brown-headed, 90, 136, 177
 call dialects, **71**
 two voices, **40**
Cracidae, 171
crane, sandhill, 5, 76, 136, 172
 female song, duets, **16**
creeper, brown, 86, 146, 175
crossbill
 Cassia, 72, 176
 red, 176
  call dialects, **72**
crow, 6, 59
 American, 136, 174
  no song?, **105**
cuckoo, 5
curlew, long-billed, 76, 172
Cyanocitta
 cristata. See jay, blue
 stelleri. See jay, Steller's

dawn singing, **112–23**
 function of, 112
 many examples, 113
development of song. *See* inborn
  song, improvised song, learned
  song
devil, Tasmanian, 170
dialects, **63–72**
dickcissel, 87, 178
 dialects, **65–66**
 polygyny and song, 30
 Why sing?, **29**
dinosaur, 5
DNA fingerprinting, 26
Dolichonyx oryzivorus. See bobolink
dove, 5, 36
 inborn song, 47
 mourning, 113, 171
  inborn song, **45**
  mechanical sounds, **21**

Dryobates
 pubescens. See woodpecker, downy
 villosus. See woodpecker, hairy
Dryocopus pileatus. See
  woodpecker, pileated
duck, 5, 170
 plumed whistling, 170
 ruddy, 171
  mechanical sounds, **22**
 wood, 7, 171
duets, female song, **15–21**
Dumetella carolinensis. See catbird,
 gray

Empidonax, 85, 113, 147
 alnorum. See flycatcher, alder
 difficilis. See flycatcher,
  Pacific-slope
 flaviventris. See flycatcher,
  yellow-bellied
 hammondii. See flycatcher,
  Hammond's
 minimus. See flycatcher, least
 oberholseri. See flycatcher, dusky
 occidentalis. See flycatcher,
  cordilleran
 traillii. See flycatcher, willow
 virescens. See flycatcher, Acadian
 wrightii. See flycatcher, gray
Eremophila alpestris. See lark, horned
evolution of song, **147–52**
exploring birdsong, 2
eyes, hearing with, 3

fairy-wren, red-backed, 170
Falco sparverius. See kestrel,
 American
Falconidae, 173
Falconiformes, 173
fantail, rufous, 170
female song, duets, **15–21**
finch
 Cassin's, 78, 176
  courtship song, **31**

finch (*cont.*)
  house, 74, 88, 104, 113, 176
    courtship song, **31**
  finch, purple, 88, 104, 146, 176
    musicality, **161**
flamboyance. *See* grackle,
    great-tailed
flicker, northern, 7, 73–77, 93, 146,
    173
  mechanical sounds, **23**
flight songs and calls, **130–36**
  many examples, 136
flycatcher, 6, 35, 44, 63, 83, 140
  Acadian, 152, 174
  alder, 85, 152, 174
    inborn song, **45–46**
    singing in brain, 36
  cordilleran, 151–52, 174
  crested, 123
  dusky, 152, 174
  Empidonax
    song evolution, **151**
  gray, 152, 174
  great crested, 74, 77, 113, 173
  Hammond's, 152, 174
  inborn song, 47
  least, 85, 93, 113, 152, 174
    song complexity, **79**
  olive-sided, 82, 113, 146, 173
    song evolution, 149
    song repertoire, **84**
  Pacific-slope, 151–52, 174
  willow, 49, 85, 113, 141, 152,
    174
    how many species?, **142**
    inborn song, **45–46**
    singing in brain, 36
  yellow-bellied, 152, 174
flycatcher, Traill's, 45
Fringillidae, 176
frog, 5
function of song, **26–30**

Galliformes, 171

Gallinago delicata. See snipe,
    Wilson's
Gallus gallus. See junglefowl, red
Gavia immer. See loon, common
Gaviidae, 172
Gaviiformes, 172
geographic variation, 9
Geothlypis, 152
  tolmiei. See warbler, MacGillivray's
  trichas. See yellowthroat, common
goldfinch, 72, 104, 133
  American, 176
    flight songs, **134**
  Lawrence's, 176
    mimicry of calls, 77
  lesser, 176
    mimicry of calls, 77
goose, 5, 170
  Canada, 136, 171
    song vs. call, **12**
grackle
  common, 169, 177
  great-tailed, 113, 177
    awaking, **109**
    musicality, **166**
    polygyny and song, 29
grebe, 5
  pied-billed, 171
    song or call?, **15**
grosbeak, 170
  black-headed, 163, 178
    dawn singing, **115**
    female song, duets, 21
  blue, 169, 178
  evening, 146, 176
    no song?, **104**
  pine, 72, 176
  rose-breasted, 115, 163, 178
    courtship song, **32**
    female song, duets, **20**, 21
grouse, 4–5, 36
  ruffed, 7, 43, 171
    how many species?, **141–142**
    mechanical sounds, 24

Gruidae, 172
Gruiformes, 172
guillemot, pigeon, 172
gull, 5, 83
    laughing, 7, 172
    western, 172
        singing in brain, **36**
Gymnorhinus cyanocephalus. See jay,
    pinyon

Haemorhous
    cassinii. See finch, Cassin's
    mexicanus. See finch, house
    purpureus. See finch, purple
hawk, 5, 10, 73
    broad-winged, 78, 173
    red-shouldered, 78, 146, 173
        song vs. call, **14**
    red-tailed, 146, 173
        flight songs, **132**
    ·   red-tailed, 7, 76, 78
headphones, using, 2
hearing, **41–43**
heron, 5
    heron, great blue, 173
        night singing, **126**
Hirundinidae, 175
Hirundo rustica. See swallow, barn
honeyeater, scarlet, 170
hummingbird, 5, 50
    Anna's, 172
        dialects, **69**
        polygyny and song, 29
    broad-tailed, 136, 172
        mechanical sounds, **22**
        polygyny and song, 29
    ruby-throated, 7, 172
hybrid songs, 45
Hylocichla mustelina. See thrush,
    wood

Icteria virens. See chat,
    yellow-breasted
Icteridae, 177

Icterus
    galbula. See oriole, Baltimore
    parisorum. See oriole, Scott's
    spurius. See oriole, orchard
improvised songs, **47–49**
inborn songs, **44–47**
    and geographic variation, 47
individuality in songs, **137–40**
Ixoreus naevius. See thrush, varied

jay, 6, 59, 73
    blue, 73–75, 105–06, 113, 174
        awaking, **110**
        call matching, 110
        mimicry, **78**
    pinyon, 174
        no song?, **105**
    Steller's, 78, 105, 174
Juliet Principle, 169
Junco hyemalis. See junco, dark-
    eyed
junco, dark-eyed, 88, 97, 113, 146,
    177
    song complexity, **81**
junglefowl, red, 171
    hearing, **43**
    individuality, **137**

kestrel, American, 132, 173
killdeer, 74, 76, 136, 172
    song vs. call, **13**
kingbird, 6, 123
    Couch's, 113, 173
        song evolution, **149**
    eastern, 7, 113, 136, 173
        song evolution, **149**
    western, 74, 113, 173
        song evolution, **149**
kingfisher, 5
    belted, 7, 74
kinglet
    golden-crowned, 169, 175
    ruby-crowned, 175
        song repertoire, **86**

Laniidae, 174
Lanius ludovicianus. See shrike,
    loggerhead
Laridae, 172
lark, horned, 136, 174
    dawn singing, **114**, **115**
Larus occidentalis. See gull, western
larynx, 38
Latin names, **170–78**
learned songs, **49–55**
learning decisions, **56–58**
Leucophaeus atricilla. See gull,
    laughing
limpkin, 172
    night singing, **125**
lion, 5
listen, how to, 1
lizard, 5
longspur, McCown's, 176
    flight songs, **134–35**
loon, 5
    common, 92, 172
        musicality, **164**
Loxia curvirostra. See crossbill, red

magpie, 74
    black-billed, 105, 174
mallard, 171
    song vs. call, **12**
Malurus melanocephalus. See fairy-
    wren, red-backed
mammal, 5
manakin, 6
Manorina melanophrys. See miner,
    bell
Mareca americana. See wigeon,
    American
martin, purple, 74, 113, 146, 175
    night singing, **126–27**
matched countersinging, 61
matching, song and call, **58–63**
meadowlark
    eastern, 92, 97, 151, 177
        song evolution, 147, **150**

western, 2, 83, 90, 97, 113, 151,
    162–63, 177
    flight songs, **135**
    repertoire, 37
    song evolution, 147, **150**
mechanical sounds, **21–25**
Megaceryle alcyon. See kingfisher,
    belted
Megascops
    asio. See screech-owl, eastern
    kennicottii. See screech-owl,
        western
Melanerpes
    carolinus. See woodpecker,
        red-bellied
    erythrocephalus. See woodpecker,
        red-headed
Meleagris gallopavo. See turkey, wild
Melospiza melodia. See sparrow,
    song
memory for song, 37
migrants, 26, 53, 54
    calling, 10
mimicry, **73–78**
    mystery of, 73, 78
mimic-thrushes, 49
Mimidae, 176
Mimus polyglottos. See mockingbird,
    northern
miner, bell, 170
mini-dialects, **56–58**
mockingbird, northern, 91, 100,
    113, 158, 176
    awaking, **111**
    estimating repertoire size of, 75
    mimicry, **73**, **75**
    night singing, **127**
    repertoire, 37
    song matching, **62**
mockingbirdese, 62, 73
Molothrus ater. See cowbird,
    brown-headed
monogamy, social, 26
mouse, 5

musicality, **153–68**
  accelerando, **154**
  contrasts, **156**
  crescendo, **157**
  diminuendo, **157**
  dissonance, **159**
  improvisation, **159**
  Messiaen, Olivier, 153
  metronome, **160**
  pitch-shifting, **155**, **156**
  Respighi's *Pines of Rome,* 153
  ritardando, **154**
  theme and variations, **158**
  Vivaldi's *Four Seasons,* 153
Myadestes townsendi. See solitaire,
  Townsend's
Myiarchus crinitus. See flycatcher,
  great crested
Myzomela sanguinolenta. See
  honeyeater, scarlet

night singing, **123–29**
nighthawk, common, 83, 136, 171
  mechanical sounds, **22**
night-heron, black-crowned, 76, 173
nightingale, 153
nightjar, 5, 28
nomadism and song improvisation,
  49
Nucifraga columbiana. See
  nutcracker, Clark's
Numenius americanus. See curlew,
  long-billed
nutcracker, Clark's, 174
nuthatch
  red-breasted, 9, 169, 175
  white-breasted, 73, 77, 81, 88, 97,
    146, 175
      song and call, **9**
Nycticorax nycticorax. See night-
  heron, black-crowned

Odontophoridae, 171
Old Faithful, 133

Oporornis, 128
  agilis. See warbler, Connecticut
Oreoscoptes montanus. See thrasher,
  sage
Oreothlypis
  peregrina. See warbler, Tennessee
  ruficapilla. See warbler, Nashville
oriole
  Baltimore, 177
    female song, duets, **18**
  orchard, 146, 177
    dawn singing, **116**
  Scott's, 162, 177
oropendola, Montezuma, 169, 177
Ortalis vetula. See chachalaca, plain
Orthonyx spaldingii. See chowchilla
osprey, 5
ovenbird, 81, 87, 116, 136, 178
  crescendo, **157**
  flight songs, **129**
  individuality, **138**
  night singing, **129**
owl, 5, 8, 36
  barred, 125, 173
    night singing, **124**
  great horned, 7, 81, 106, 125, 173
  northern hawk, 10, 173
owling, **125**
Oxyura jamaicensis. See duck, ruddy

Pachycephala
  rufiventris. See whistler, rufous
  pectoralis. See whistler, golden
packages, song, **101**
Paridae, 175
Parkesia noveboracensis. See
  waterthrush, northern
parrot, 5, 50
Parulidae, 178
Passer domesticus. See sparrow, house
Passerculus sandwichensis. See
  sparrow, savannah
Passerella iliaca. See sparrow, fox
Passerellidae, 176

Passeridae, 176
Passeriformes, 173
Passerina, 152
    amoena. See bunting, lazuli
    caerulea. See grosbeak, blue
    ciris. See bunting, painted
    cyanea. See bunting, indigo
passerine, 5, 35
Pelecaniformes, 172
perching birds, 5
Peterson, Roger Tory, 41, 161
Petrochelidon pyrrhonota. See
    swallow, cliff
Peucaea
    aestivalis. See sparrow, Bachman's
    cassinii. See sparrow, Cassin's
Phalacrocoracidae, 172
Phalacrocorax auritus. See cormorant,
    double-crested
Phalaenoptilus nuttallii. See poorwill,
    common
Phasianidae, 171
Pheucticus
    ludovicianus. See grosbeak,
        rose-breasted
    melanocephalus. See grosbeak,
        black-headed
phoebe, 6, 123
    black, 85, 113, 152, 174
        song repertoire, **84**
    eastern, 73–76, 84–85, 113, 152,
        174
        inborn song, **46**, 47
        singing in brain, **35**
    Say's, 85, 113, 152, 174
        song repertoire, **84**
    song use, **85**
Pica. See magpie
    hudsonica. See magpie, black-billed
Picidae, 173
Piciformes, 173
pigeon
    homing
        hearing, 42

    rock, 7, 171
        hearing, **42**
Pinicola enucleator. See grosbeak,
    pine
Pipilo
    chlorurus. See towhee, green-tailed
    erythrophthalmus. See towhee,
        eastern
    maculatus. See towhee, spotted
Piranga
    flava. See tanager, hepatic
    ludoviciana. See tanager, western
    olivacea. See tanager, scarlet
    rubra. See tanager, summer
    bidentata. See tanager,
        flame-colored
Podicipedidae, 171
Podicipediformes, 171
Podilymbus podiceps. See grebe,
    pied-billed
Poecile, 152
    atricapillus. See chickadee,
        black-capped
    carolinensis. See chickadee,
        Carolina
    gambeli. See chickadee, mountain
    hudsonicus. See chickadee, boreal
    rufescens. See chickadee,
        chestnut-backed
polygyny and song, **29–30**
Pooecetes gramineus. See sparrow,
    vesper
poorwill, common, 171
    Why sing?, 28
Porzana carolina. See sora
prairie flight songs, exploring, 135
prairie-chickens, 36
Procellariidae, 172
Procellariiformes, 172
Procnias spp. *See* bellbird
Progne subis. See martin, purple
Psaltriparus minimus. See bushtit
Psarocolius montezuma. See
    oropendola, Montezuma

Psophodes olivaceus. See whipbird, eastern
Pyrrhuloxia, 97, 178
  song matching, **59**, **60**
  song repertoire, 89

QR codes, 2
quail, 36
  inborn song, 47
  California, 76, 171
    singing in brain, **36**
questions, asking, 1
Quiscalus
  mexicanus. See grackle, great-tailed
  quiscula. See grackle, common

rail, 5
  Virginia, 172
    Why sing?, **27**
Rallidae, 172
Rallus limicola. See rail, Virginia
Raven Lite, 3, 40, 41, 43, 48, 69, 75, 89, 101, 115, 122, 130, 137, 178
raven
  common, 105, 136, 146, 174
    call matching, **59**
  forest, 170
recording birdsong, 179
redpoll, common, 176
  flight songs, **133**
redstart, American, 88, 113, 116, 158, 178
  individuality, **139**
  subsong, **53**
Regulidae, 175
Regulus
  calendula. See kinglet, ruby-crowned
  satrapa. See kinglet, golden-crowned
relationships among species, **170–78**
repertoire delivery, **94–103**
repertoire sizes, **83–94**

reptile, 5
resources, additional, 179
Rhipidura
  leucophrys. See wagtail, willie
  rufifrons. See fantail, rufous
Rhynchophanes mccownii. See longspur, McCown's
robin, American, 1–3, 7, 74–76, 103, 106, 113, 176
  repertoire delivery, **101**
  song and call, **10**, 11
rooster, 29
roosting and awaking, **108–12**

Salpinctes obsoletus. See wren, rock
sandpiper, upland, 172
  musicality, **163**
sapsucker
  mechanical sounds, **24–25**
  red-breasted, 24–25, 152, 173
  red-naped, 24–25, 152, 173
  yellow-bellied, 24–25, 113, 152, 173
Sarcophilus harrisii. See devil, Tasmanian
Sayornis, 152
  nigricans. See phoebe, black
  phoebe. See phoebe, eastern
  saya. See phoebe, Say's
Scenopoeetes dentirostris. See bowerbird, tooth-billed
science
  definition, 4
  doing, 4
scientific names, **170–78**
Scolopacidae, 172
Scolopax minor. See woodcock, American
screech-owl
  eastern, 125, 173
  western, 125, 173
scrub-jay, California, 62, 106, 174
seeing birdsong, 178
Seiurus aurocapilla. See ovenbird

Selasphorus platycercus. *See*
    hummingbird, broad-tailed
Setophaga, 147, 152
    caerulescens. *See* warbler, black-
        throated blue
    magnolia. *See* warbler, magnolia
    pensylvanica. *See* warbler,
        chestnut-sided
    petechia. *See* warbler, yellow
    ruticilla. *See* redstart, American
shearwater, 5
    sooty, 172
            female song, duets, **17**
shorebird, 5, 163
shrike, loggerhead, 81, 97, 174
    metronome, **160**
Sialia
    currucoides. *See* bluebird,
        mountain
    sialis. *See* bluebird, eastern
Sitta
    canadensis. *See* nuthatch,
        red-breasted
    carolinensis. *See* nuthatch,
        white-breasted
Sittidae, 175
skylark, 153
snipe, 4, 76, 136, 172
    Wilson's
            flight songs, 130
            mechanical sounds, **23**
solitaire, Townsend's, 43, 102, 163,
    176
    song repertoire, 91
    song complexity, **82**
song control centers, 38
song vs. call, **6–15**
songbird, defined, 6
songless songbirds, **104–7**
sora, 83, 172
    Why sing?, **27**
sparrow, 6
    American tree, 86, 177
        dialects, **66**

Bachman's, 113, 163, 177
    repertoire delivery, **96**
black-throated, 113, 177
    repertoire delivery, **95**
Brewer's, 83, 152, 177
    dawn singing, **121**, **122**
Cassin's, 88, 136, 163, 177
    repertoire delivery, **96**
chipping, 81, 86, 93, 121, 152,
    177
        dawn singing, **120**
        dialects, **57**, 63, 140
        song learning decisions,
            56–57
clay-colored, 81, 152, 177
    dawn singing, **122**
    explore!, **123**
field, 88, 152, 177
    accelerando, **154**
    dawn singing, **121**
fox, 88, 100, 146, 163, 177
    how many species?, **145**
    sooty, 66, 145
    slate-colored, 66, 145
    red, **66**, 145
    large-billed, 66, 145
golden-crowned, 169, 177
Henslow's, 86, 177
    hearing, **41**
house, 74–76, 176
    song complexity, **80**
lark, 177
    courtship song, **32**
olive, 169, 176
rufous-crowned, 169, 177
savannah, 146, 177
    song repertoire, **86**
seaside, 169, 177
song, 37, 90, 97, 177
    subsong, 50, **51**
vesper, 113, 177
    dialects, **67**
white-crowned, 87, 146, 152, 177
    dialects, **64**, 140

white-throated, 87, 152, 158,
163, 177
dialects, **67**, **68**
subsong, **52**, **54**
species conundrums, list of, **146**
species, recognizing, **140–46**
species, searching for new, 142
Sphyrapicus, 152
nuchalis. See sapsucker, red-naped
ruber. See sapsucker, red-breasted
varius. See sapsucker,
yellow-bellied
Spinus
lawrencei. See goldfinch,
Lawrence's
psaltria. See goldfinch, lesser
tristis. See goldfinch, American
Spiza americana. See dickcissel
Spizella, 152
breweri. See sparrow, Brewer's
pallida. See sparrow, clay-colored
passerina. See sparrow, chipping
pusilla. See sparrow, field
species comparisons, **120**
Spizelloides arborea. See sparrow,
American tree
starling, European, 176
mimicry, **76**
Streptopelia decaocto. See collared-
dove, Eurasian
Strigidae, 173
Strigiformes, 173
Strix varia. See owl, barred
Sturnella
magna. See meadowlark, eastern
neglecta. See meadowlark, western
Sturnidae, 176
Sturnus vulgaris. See starling,
European
suboscine, 35, 44
defined, 6
subsong. *See* babbling
Suliformes, 172
Surnia ulula. See owl, northern hawk

swallow, 6, 74
barn, 175
musicality, **165**
cliff, 175
musicality, **166**
tree, 76, 88, 113, 175
explore, 133
flight songs, **132**
swift, 5, 36, 136, 171
chimney
song vs. call, **13**
Sylviidae, 175
syringes (voice boxes), 33, **38–41**
anatomy of, 38

Tachycineta bicolor. See swallow, tree
tanager
flame-colored, 148, 178
hepatic, 148, 178
scarlet, 3, 94, 113, 178
song evolution, **147–48**
summer, 77, 113, 178
song evolution, **147–48**
western, 113,178
song evolution, **147–48**
taxonomy of species, **170–78**
temporal resolving power, 41
thrasher, 49, 73
brown, 2, 74, 100, 113, 152, 176
improvised song, 49
mimicry, 78, **92**
song repertoire, **33**, 37,
91–94
theme and variations, **158**
vocal matching, **61**
Why sing?, **28**, 30
California, 113, 152, 176
mimicry, 78
singing in brain, **33–34**
sage, 2, 102, 113, 152, 163, 176
mimicry, 78
repertoire, 37
singing in brain, **37**
song complexity, **83**

thrasher (*cont.*)
    song repertoire, 91
thrush, 1, 6
    gray-cheeked, 88, 100, 152, 162, 176
        night singing, **128**
    hermit, 100, 146, 152, 157–58, 163, 176
        contrasts, **156**
        song magnificence, **89**
        song repertoire, **88**
    song, 153
    Swainson's, 43, 88, 95, 100, 113, 146, 152, 176
        awaking, **110, 111**
        call matching, 110
        musicality, **161**
        night singing, 128
    varied, 43, 100, 176
        dissonance, **159**
        two voices, **39**
    wood, 43, 73–75, 77, 100, 158, 163, 176
        repertoire delivery, **102**
        roosting, **109**
        two voices, **38–39**
Thryomanes bewickii. See wren, Bewick's
Thryothorus ludovicianus. See wren, Carolina
titmouse
    black-crested, 169, 175
    juniper, 175
        song matching, 63
    oak, 81, 90, 97, 175
        song matching, **63**
    tufted, 73–74, 81, 97, 175
        song matching, 63
        song repertoire, **87**
toad, 5
towhee
    eastern, 87–88, 90, 95, 97, 113, 177
        repertoire delivery, **94**

    green-tailed, 100, 146, 177
    spotted, 65, 90, 95, 97, 177
        repertoire delivery, **94**
Toxostoma, 152
    redivivum. See thrasher, California
    rufum. See thrasher, brown
tree-frog, Pacific, 76
Tringa semipalmata. See willet
Trochilidae, 172
Troglodytes
    aedon. See wren, house
    hiemalis. See wren, winter
    pacificus. See wren, Pacific
Troglodytidae, 175
Turdidae, 176
Turdus
    merula. See blackbird, Eurasian
    migratorius. See robin, American
turkey, wild, 36, 171
    Why sing?, **29**
Tyrannidae, 173
Tyrannus, 149
    couchii. See kingbird, Couch's
    tyrannus. See kingbird, eastern
    verticalis. See kingbird, western

veery, 43, 88, 100, 113, 152, 176
    awaking, **110, 111**
    call matching, 110
    song and call, **10**
vireo
    Bell's, 90, 113, 146, 152, 174
        musicality, **164**
    black-capped, 152, 169, 174
    blue-headed, 152, 169, 174
    Cassin's, 152, 174
    Hutton's, 152, 174
    Philadelphia, 88, 152, 174
        repertoire delivery, **98–99**
    plumbeous, 152, 174
    red-eyed, 152, 174
        improvised song, **48**
        repertoire delivery, **98–99**
        subsong, **50**

warbling, 141, 152, 174
    how many species?, **143**
    subsong, **50**
white-eyed, 90, 152, 174
    mimicry of calls, **77**
yellow-throated, 152, 174
    repertoire delivery, **101**
Vireo, 98, 152
    atricapilla. See vireo, black-capped
    bellii. See vireo, Bell's
    cassinii. See vireo, Cassin's
    flavifrons. See vireo,
        yellow-throated
    gilvus. See vireo, warbling
    griseus. See vireo, white-eyed
    huttoni. See vireo, Hutton's
    olivaceus. See vireo, red-eyed
    philadelphicus. See vireo,
        Philadelphia
    plumbeus. See vireo, plumbeous
    solitarius. See vireo, blue-headed
Vireonidae, 174
voice boxes. *See* syringes

wagtail, willie, 170
warbler, 6, 147, 163
    black-throated blue, 88, 116, 152,
      178
        why sing?, **28**, 30
    chestnut-sided, 88, 113, 116, 152,
      178
        individuality, **139**
    Connecticut, 87, 178
        night singing, **128**
    dawn singing, **116**
    MacGillivray's, 152, 178
    magnolia, 87, 116, 152, 178
        subsong, **54**
    Nashville, 93, 178
    red-faced, 162, 178
    Tennessee, 87, 116, 178
        subsong, **53**
    yellow, 90, 116, 146, 152, 162, 178
        dawn singing, **115**

waterthrush, northern, 87, 178
waxwing, cedar, 176
    no song?, **104**
website for the book, 2
whipbird, eastern, 170
whip-poor-will, 28
    eastern, 152, 171
        inborn song, 44
        Why sing?, **27**
    Mexican, 7, 152, 171
        inborn song, 44
        Why sing?, **27**
whistler
    golden, 170
    rufous, 170
wigeon, American, 171
    song or call?, **15**
willet, 172
    flight songs, **131**
wonder in birdsong, 1
woodcock, American, 172
    explore, 131
    flight songs, **130**
    polygyny and song, 29
woodcreeper, 6
woodpecker, 4, 5
    brains of, 23
    downy, 76, 173
    hairy, 76, 173
    pileated, 173
        mechanical sounds, **24**
        roosting, **108**
    red-bellied, 75, 173
        song or call?, **15**
    red-headed, 74, 173
wood-pewee, 6, 123
    eastern, 76, 84–85, 92, 93, 113,
      163, 174
        song evolution, **148–49**
    western, 84–85, 113, 173
        song evolution, **148–49**
wren, 1, 6, 15
    Bewick's, 90, 97, 113, 146, 163,
      168, 175

wren (*cont.*)

    how many species?, **144**

    musicality, **167**

cactus, 97, 127, 175

    song complexity, **80**

canyon, 97, 163, 175

    ritardando, **154**

Carolina, 73–75, 77, 81, 97, 175

    female song, duets, **17**

    singing in brain, **35**

    subsong, 50, **51**, **55**

eastern marsh, 140, 145

    singing in brain, **34**

house, 175

    song complexity, **82**

marsh, 113, 146, 152, 175

    learned song, 48

    polygyny and song, 29, 30

Pacific, 90, 97, 175

    hearing, **42**

    musicality, 167

rock, 74, 81, 113, 175

    song repertoire, **90**

sedge, 97, 152, 175

    how many species?, **144**

    improvisation, 48, **159**

    polygyny and song, 29

    song repertoire, **91**

western marsh, 90–91, 100, 140

    singing in brain, **34**

    song matching, **60**

    vocal matching, **61**

winter, 88, 97, 175

    musicality, **166**

wrentit, 86, 175

    female song, duets, **18**

Xanthocephalus xanthocephalus. See
    blackbird, yellow-headed

yellowstart, 139

yellowthroat, common, 73–74, 81,
    87, 116, 146, 152, 178

    dialects, 58, 63, 140

    song and call, **11**

    song learning decisions, **58**

Zenaida macroura. See dove,
    mourning

Zonotrichia, 152

    albicollis. See sparrow,
      white-throated

    atricapilla. See sparrow,
      golden-crowned

    leucophrys. See sparrow,
      white-crowned